ARTIFICIAL EARTH

Before you start to read this book, take this moment to think about making a donation to punctum books, an independent non-profit press,

@ https://punctumbooks.com/support/

If you're reading the e-book, you can click on the image below to go directly to our donations site. Any amount, no matter the size, is appreciated and will help us to keep our ship of fools afloat. Contributions from dedicated readers will also help us to keep our commons open and to cultivate new work that can't find a welcoming port elsewhere. Our adventure is not possible without your support.

Vive la Open Access.

Fig. 1. Detail from Hieronymus Bosch, *Ship of Fools* (1490–1500)

First published in 2023 by punctum books, Earth, Milky Way.
https://punctumbooks.com

ISBN-13: 978-1-68571-130-6 (print)
ISBN-13: 978-1-68571-131-3 (ePDF)

DOI: 10.53288/0406.1.00

LCCN: 2023947842
Library of Congress Cataloging Data is available from the Library of Congress

Book design: Hatim Eujayl and Vincent W.J. van Gerven Oei
Cover image: Konstantin Yuon, *New Planet* (1921).

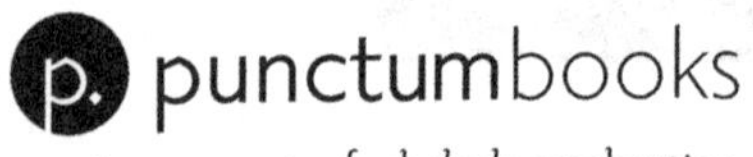

HIC SVNT MONSTRA

J. Daniel Andersson

ARTIFICIAL EARTH

A Genealogy of Planetary Technicity

p.

Contents

Acknowledgments

It is sometimes said that writing a book is a lonely enterprise. In this case, nothing could be further from the truth. *Artificial Earth: A Genealogy of Planetary Technicity* was written under the auspices of substantial personal and institutional support. To begin with, the Department of Thematic Studies at Linköping University has provided precisely the kind of friendly but rigorous atmosphere conducive to making the whole process into a genuine intellectual adventure. This includes my wonderful colleagues at the Seed Box: A Mistra-Formas Environmental Humanities Collaboratory; the Center for Climate Science and Policy Research; and the Linköping University Negative Emission Technologies network. Moreover, I have benefited from invitations to present my work in progress at the Department of Sociology at Uppsala University, the Department of Urban Planning and Environment and the Environmental Humanities Laboratory at the KTH Royal Institute of Technology, and at the Forum Scientiarum at Karl Eberhards University of Tübingen. Various connections through the School of Environmental Studies at Queen's University, the Department of Gender and Cultural Studies at the University of Sydney, and the Department of Media, Communications, and Cultural Studies at Goldsmiths have also played important roles in shaping an unruly manuscript into a finished product.

On a personal level, I want to begin by expressing my deepest appreciation for the intellectual generosity, mentoring, and friendship of Jonas Anshelm, Johan Hedrén, and Simon Haikola. The same can be said about Anders Hansson and Mathias Fridahl, who welcomed me into their ranks at a critical stage in the completion of this book. Without their faith in me, the manuscript would still remain unfinished. As for critical feedback, I am particularly indebted to Sabine Höhler, Eva Lövbrand, and Alf Hornborg. Many thanks for support throughout various stages of the journey are extended also to Kristin Zeiler, Björn-Ola Linnér, Steve Woolgar, Ola Uhrqvist, Cecilia Åsberg, Tora Holmberg, Karin Bradley, Jesper Olsson, Victoria Wibeck, Per Gyberg, Christer Nordlund, Carin Franzén, Myra Hird, Astrida Neimanis, Matthew Fuller, Olga Cielemęcka, Serpil Opperman, Erik Malmqvist, Lauren LaFauci, Ann-Sofie Kall, and Bruno Latour. Countless others have offered less formal but no less meaningful forms of friendship along the way: Fredrik Envall, Åsa Callmer, María Langa, Justin Makii, Martin Hultman, James Wilkes, Jonas Müller, Anna Kaijser, Johan Gärdebo, Irma Allen, Jesse Peterson, Daniele Valisena, Nimmo Elmi, Amelia Mutter, Jeffrey Christensen, Stephanie Erev, Alejandra Ruales, Emelie Fälton, Maria Jernnäs, and Lene Asp Frederiksen, in particular. Outside of academia, Catharina Andersson has been my rock this entire time. I cannot begin to convey my gratitude for her endless love and affection.

Last but not least, I want to thank Eileen A. Fradenburg Joy, Vincent W. J. van Gerven Oei, and the good people at punctum books for taking a chance on my proposal and for providing excellent support throughout the entire publication process. Thanks also to the associated peer reviewers for encouragement and advice.

I like to think

(it has to be!)

of a cybernetic ecology

where we are free of our labors

and joined back to nature,

returned to our mammal

brothers and sisters,

and all watched over

by machines of loving grace.

— Richard Brautigan,
"All Watched Over by Machines of Loving Grace" (1967)

INTRODUCTION

The Wholly Enlightened Earth...

We do not encounter the earth under conditions of our own choosing, and its revelation is not under our control. Such is the animating principle of *Artificial Earth*. At a first glance, this might seem like a strange starting point for something written in the midst of a climate emergency, with a rising global average temperature, shrinking ice sheets, an increased frequency of extreme weather events, and degradation of ecosystem services. Since at least after World War II, the emergence of humanity as a geological agent has begun to register on a global level, which is to say that the earth is symptomatically expressing the effects of collective human activity on a planetary scale. In the context of either a catastrophic or optimistic narrative, it seems all but certain that human agency has never played a more fundamental role in deciding the future of our planet. Surely, if dangerous climate change and serious ecological harm are to be avoided, then it is precisely a question of what we as humans choose to do, based upon an assessment of the best available research. Yet, if the essential aim of technological intervention into the natural world has been to bring it under human supervision, then, as our current predicament shows, it has in practice only resulted in leaving it less controllable. Hitherto taken for granted as an immutable background for human flourishing, our planet's geospheres have through various ecological, climatological, and other global environmental crises begun to reemerge front and center, defiantly striking back at the very heart of Western society's techno-industrial hubris.

Let us take the words of the atmospheric chemist and Nobel laureate Paul Crutzen as an example. Summarizing the biogeochemical evidence for how industrial civilization has radically and permanently disrupted our planet's carbon and nitrogen cycles, ocean chemistry, and biodiversity — each one the product of millions of years of evolution — Crutzen concluded that, in terms of our scientific and technological mastery of the earth, "we are still largely treading on *terra incognita*."[1] Such an emphasis on the ontological alterity of the earth is arguably a psychoanalytic gesture. It seeks to perform a defamiliarization of the most familiar thing of all — the archetypally Freudian notion of Mother Earth, the collective home for every being, each according to its naturally endowed role. Indeed, it was Sigmund Freud who famously defined phenomena experienced as familiar yet at the same time foreign as "uncanny." The German word for the same experience, *Unheimlich*, which literally translates into "unhomely," captures even better the paradoxical notion that, according to Freud, our most haunting experiences of otherness indicate that the alien is most cleverly concealed at home. Or, as the second constellation of meaning of *Unheimlich* would have it: concealment is greatest where common sense tells us that everything has already been fully enlightened. Put differently, nearness does not mean obviousness, as was, for instance, made apparent at the moment that global warming became thinkable. Computational power allowed us to conceive of phenomena beyond the grasp of quotidian experience, but it did not so much integrate them into the emphatic dramaturgy of narrative temporality as it opened up a whole new fractal dimension of complexly bounded levels of reality. Now that the dust has settled after an intense period of globalization, and reason has shone its illuminative light on the last dark corners of the world,

---

1 Paul J. Crutzen, "Geology of Mankind," *Nature* 415, no. 6867 (2002): 23. Unless otherwise stated, all emphases in quoted material are original.

enlightenment seems only to have exploded the notion of existence as an all-inclusive receptacle into a plurality of multiple perspectives and scalar shifts. Rather than reassuringly holistic and harmoniously universal, the global phenomena of the twenty-first century point toward a fragmentary assortment of systems operating in disjointed concert—connecting, by way of weirds loops, the microscopic worlds of algae, bacteria, and viruses to the mesoscopic worlds of aquatic ecosystems, international travel, and global agriculture, all the way up to the macroscopic worlds of ocean food webs, atmospheric greenhouse gas concentrations, and global carbon and nutrients cycles. As it turns out, we live enfolded by more timescales than we can grasp.[2] With an accelerated modernity, the cumulative effect of individual lives suddenly jeopardizes the well-being of future generations, the pace of technological innovation threatens to alter the course of natural evolution, and the march of human history proves uncontainable even by the perennial rhythms of geological time. But conversely, mundane actions, when aggregated, also linger in an eerie way, as a presence felt only indirectly through, for instance, the uncanny rift between the familiar experiences of weather and the statistics of climate. Once the freak event of an unusually warm summer starts recurring, it points toward something more than a mere coincidence, yet the spectral nature of long-term averages is such that we cannot directly perceive climatological hazards, but only learn to discern their traces. On the one hand, then, conditions for life negotiated over millions of years are currently being undone in comparatively the blink of an eye, but, on the other hand, seemingly innocent everyday behavior is now capable of leaving imprints that will continue to haunt the earth for the foreseeable future and beyond.

---

2 Timothy Morton, *Dark Ecology: For a Logic of Future Coexistence* (New York: Columbia University Press, 2016), 25.

If the uncanny is the name for an "[in]between that is tainted with strangeness"[3] — that is, a disturbance to the natural order or to the customary separation of phenomena to appear within the confines of traditional registers — then it is surely the experience par excellence of global environmental change.[4] In the wake of the rapid and thorough industrialization of the nineteenth and twentieth centuries, the whole surface of the earth has to some degree been manipulated by humankind, from tropical rainforests to arctic tundra and polar icecaps, including the oceans and the atmosphere that we breathe. There is little left on our planet that humans have not, at least indirectly, left their anthropogenic fingerprints on. Yet, far from making us earthmasters, modernity has conjured into existence nonhuman forces that the Enlightenment prophets of a disenchanted nature had long since declared to be dead. For not only has nature been infused by human agency on a planetary scale, but it has been so in ways that have produced new forms of more-than-human unpredictability. Anthropogenic climate change is a great example of a radically human-caused but at the same time potentially self-amplifying runaway process, revealing the maternal security of our homely dwelling as uncannily monstrous. Without leaving our earthbound home, we have nevertheless been thrust into an unknown territory strewn with positive feedback loops of cascading effects that would threaten to catapult the planet into a hothouse state, one fundamentally at odds with the continuity of modern civilization as we know it. In the hockey-stick graphs of the "Great Acceleration,"[5] the modern promise of progress has jarringly morphed into biospheric degradation. Although nature has been seemingly denaturalized, it appears stranger than ever, and the more we shape the earth in our own image, the more foreign it seems to become.

---

3 Hélène Cixous, "Fiction and Its Phantoms: A Reading of Freud's *Das Unheimliche* (the 'Uncanny')," *New Literary History* 7, no. 3 (1976): 543.

4 Franklin Ginn et al. "Introduction: Unexpected Encounters with Deep Time," *Environmental Humanities* 10, no. 1 (2018): 213–25.

5 Will Steffen et al. "The Trajectory of the Anthropocene: The Great Acceleration," *The Anthropocene Review* 2, no. 1 (2015): 81–98.

Although it has an intellectual history of its own, the geophilosophical specificity of this experience is a relatively novel phenomenon. As suggested by the literary theorist Fredric Jameson, its specificity is contingent upon having lived through the apocalypse associated with the "end of nature"[6] — the recognition that humans are altering the earth's geospheres to the point that we can perceive our species as a global force of nature, with the consequence that human agency can no longer be approached as though it belongs to a domain apart from its ecological, mineral, chemical, and atmospheric contexts — which, he argues, even the twentieth century's most perceptive critics of technology failed to properly appreciate:

> Even Heidegger continues to entertain a phantasmatic relationship with some organic precapitalist peasant landscape and village society, which is the final form of the image of Nature in our own time. Today, however, it may be possible to think all this in a different way, at the moment of a radical eclipse of Nature itself: Heidegger's "field path" is, after all, irredeemably and irrevocably destroyed by late capital, by the green revolution, by neocolonialism and the megalopolis, which runs its superhighways over the older fields and vacant lots and turns Heidegger's "house of being" into condominiums, if not the most miserable unheated, rat-infested tenement buildings. The *other* of our society is in that sense no longer Nature at all, as it was in precapitalist societies, but something else which we must now identify.[7]

During the time that Edmund Burke wrote, for instance, nature was still feared and admired in equal measure because of humanity's seeming inability to control its forces. Well into the second half of the eighteenth century, the affective registers

6 Bill McKibben, *The End of Nature* (New York: Random House, 1989), 47. See also Paul Wapner, *Living Through the End of Nature: The Future of American Environmentalism* (Cambridge: MIT Press, 2010).

7 Fredric Jameson, *Postmodernism, or, the Cultural Logic of Late Capitalism* (Durham: Duke University Press, 1991), 34–35.

inspired by the nascent science of geology produced simultaneously terror and delight. With the rapid industrialization of the late eighteenth and early nineteenth century, however, nature's preeminent status as a limit to and condition for human flourishing had not only begun to wane but was being completely reimagined. As Immanuel Kant would later insist in a rejoinder to Burke, what begins in pain and humiliation, as the puniness and vulnerability of the human body in nature is exposed, ends in satisfying self-admiration insofar as we, through reason, "may become conscious of our superiority over nature within, and thus also over nature without us."[8] Conveniently enough, the Enlightenment promise of progress lifted the rational modern subject right out of nature by "regarding [humanity's] vocation as sublimely exalted above it."[9] It came with guarantees of taming and managing — we could even say intellectually administering — the awesome power, scale, and physical threat of nature, and all the while it asserted human entitlement to rational supremacy over the nonhuman, and over the irrational, wherever it may appear. Admittedly, the vast size and violent force of nature may put the imagination into painful crisis — and this is the moment of terror and overpowering — but reason eventually comes to the spectator's rescue by recognizing itself as a power separate from and ostensibly superior to nature, and thus, reminded of the supersensible destiny of rational moral agency, the spectator may recover its dignity. But if nature qua threat was exemplified by the 1755 Lisbon earthquake, a deadly disaster whose sensibilities reverberated across the European continent, then the late twentieth century arguably marks an end to the modern concept of the sublime by implicating us all in the planetary-wide purview of global environmental change. Peculiar about our contemporary condition of global change is that there is no longer any place or position of security from which this spectacle of terror can safely be overseen. In compar-

8 Immanuel Kant, *Critique of Judgement,* trans. James C. Meredith, ed. Nicholas Walker (Oxford: Oxford University Press, 2007), 94.
9 Ibid.

ison, violent natural hazards, such as earthquakes or tsunamis, though terrible and overpowering in their immediacy, are local enough that their consequences can be enjoyed from a distance. Although the Lisbon earthquake once shook the human imagination to its core, it was still limited — both geographically and ontologically — in such a way as to sanction a gap between the rational subject and the hazards of nature. Instead, if rational critical reflection reveals anything today, it is that the environmental risks of the twenty-first century are first and foremost manufactured, and thereby fundamentally include us.[10] In an ironic twist of fate, our technological systems of anticipation and preemption are now so sophisticated that "our cognitive powers become self-defeating. The more we know about radiation, global warming, and the other massive objects that show upon our radar, the more enmeshed in them we realize we are. Knowledge is no longer able to achieve escape velocity from Earth."[11] All positions of relative advantage ultimately vanish in the wake of the global impact of human activity.

Burke could conceive of the sublime as a failure of human artifice to ever measure up to the overwhelming power of the natural world, but today nature has all but seemingly been conquered by artifice instead. If anything, it is no longer nature that is sublime, but rather "that enormous properly human and antinatural power of dead human labor stored up in our machinery — an alienated power [...] which turns back on and against us in unrecognizable forms and seems to constitute the massive dystopian horizon of our collective as well as our individual praxis."[12] It is our own global technological infrastructures that now exceed our cognitive powers of representation and calculation — and our practical capacities of manipulation and control.

---

10 Gene Ray, "Terror and the Sublime in the So-Called Anthropocene," *Liminalities: A Journal of Performance Studies* 16, no. 2 (2020): 3–4.

11 Timothy Morton, *Hyperobjects: Philosophy and Ecology after the End of the World* (Minneapolis: University of Minnesota Press, 2013), 160.

12 Jameson, *Postmodernism*, 35. See also Leo Marx, *The Machine in the Garden: Technology and the Pastoral Ideal in America* (Oxford: Oxford University Press, 2000), 195.

… Is Radiant with Triumphant Calamity

As a response to the unhomely experience of living in the midst of these global environmental changes, such a hysterically sublime perpetuation of existential unease has been a frequent theme within the literary genre of cyberpunk.[13] Faced with a condition that has made the traditional position of the active and knowing subject, ontologically separated from a passive and objective world, look increasingly untenable, these sci-fi writers have attempted to find new ways to imagine our (immanent) relation to the immense architecture of globalization, and "one of the most popular means of representing this relation has been to figure the human subject as immersed in a vast and inescapably complex, technological space."[14] It is this figuration that links, for instance, Jameson's theorizing of the subject's bewildering absorption in hyperspace to the dense noir visuals in such films as *Blade Runner* or to the fluidity between interior and exterior space in the Sprawl trilogy of William Gibson. In both cases, the (dis)organization of space is presented as an "alarming disjunction point between the body and its environment," a cybernetic ecosystem of decentered networks "in which we find ourselves caught as individual subjects,"[15] what Jameson describes as an ego-shattering experience of disorientation. "This latest mutation in space," he writes, "has finally succeeded in transcending the capacities of the individual human body to locate itself, to organize its immediate surroundings perceptually, and cognitively to map its position in a mappable external world."[16] In both cases too, it is viewed as explicitly technological. Not in the sense of a modern aesthetic of industrial machines — which is present only as a pastiche of past styles — but in the postmod-

13 Jameson, *Postmodernism*, 38. See also Scott Bukatman, *Terminal Identity: The Virtual Subject in Postmodern Science Fiction* (Durham: Duke University Press, 1993).

14 R.L. Rutsky, *High Technē: Art and Technology from the Machine Aesthetic to the Posthuman* (Minneapolis: University of Minnesota Press, 1999), 14.

15 Jameson, *Postmodernism*, 44.

16 Ibid.

ern sense of a synthetic environment of machinic assemblages: an ontologically flat space of surfaces, images, simulations, and empty signifiers.[17] So, although the title of this book refers to an "artificial earth," the associations it wishes to evoke are not those of a complete domestication of nature. Quite to the contrary, of central concern is the way in which the radical eclipse of nature "liquidates all internal moments of enjoyment and ends, not in self-admiration, but in shame, shudder, and deeper subjective crisis."[18] Insofar as it is inspired by cyberpunk, *Artificial Earth* is less interested in its dystopian visuals of a near-future megalopolis — with its endless urban jungle of dilapidated apartment complexes, smoke-spewing factories, and neon-decorated high-rises — than in the persistent dedication of its writer to explore uncanny forms of alienation associated with a world that has been entirely humanized, such as the uncomfortable impression that its inhabitants have correspondingly lost a firm sense of their own humanity.

In Philip K. Dick's sci-fi juxtaposition of high-tech society and biospheric collapse, such an eclipse has been presaged as a dire expression of Theodor Adorno and Max Horkheimer's declaration, on the opening page of their *Dialectic of Enlightenment* (1944), that "the wholly enlightened earth is radiant with triumphant calamity."[19] In both cases, the technological triumph of humankind is starkly contrasted with its spiritual defeat. Following the instructions of Baconian science, humans began replacing their spiritual connection to nature with a physical one, but now, as nature has been successfully subdued, reduced to nothing but the stimulus response of its most basic elements, the last artifacts of nature's existence, humans too find themselves subjugated to the same instrumental impetus of being treated as a means rather than an end. In order to grasp this contradiction, whereby a completely enlightened earth had led

17 Ibid., 385.

18 Ray, "Terror and the Sublime," 5.

19 Max Horkheimer and Theodor W. Adorno, *Dialectic of Enlightenment: Philosophical Fragments,* ed. Gunzelin Schmid Noerr, trans. Edmund Jephcott (Stanford: Stanford University Press, 2002), 1.

only to the reification of the human, Horkheimer and Adorno suggested that we must conceive of humanity's instrumental domination of nature in a dialectical fashion. Such an approach mirrors Heidegger's observation that the ordering of modernity's instrumentalizing impetus is essentially disordering, and its orienting essentially disorienting. The translator Samuel Weber has trenchantly pointed out that "although [the English rendition] takes the collecting, assembling function of the *Gestell* into account, it effaces the tension between verb and noun that resounds in the German and that points to the strange, indeed uncanny, mixture of movement and stasis that distinguishes the goings-on of modern technics and upon which Heidegger places considerable emphasis."[20] Paradoxically, "the more technics seeks *to place* the subject into safety, the less safe its *places* become. The more it seeks to place its orders, the less orderly are its emplacements."[21] In Heidegger's diagnosis, then, the conscious exploitation of nature is inextricably interlinked with the unconscious reification of the human: "In the planetary imperialism of technologically organized man, the subjectivism of man attains its acme, from which point it will descend to the level of organized uniformity and there firmly establish itself. This uniformity becomes the surest instrument of total, i.e., technological, rule over the earth."[22] It is in enframing—which challenges forth the entire earth as a standing reserve and thereby dispossess the human too of any other place to stand except as a stockpiled bystander, on standby as an abstract numeral qua productivity to be administered, regulated, and managed much like any other resource—that humans are instrumentalized into beings that order without asking questions, that objectify the world around them, and that consequently abandon any real

20 Samuel Weber, "Upsetting the Setup: Remarks on Heidegger's Questing after Technics," in *Mass Mediauras: Form, Technics, Media,* ed. Alan Cholodenko (Stanford: Stanford University Press, 1996), 71.

21 Ibid., 74.

22 Martin Heidegger, "The Age of the World Picture," in *The Question concerning Technology and Other Essays,* trans. William Lovitt (New York: Harper & Row, 1977), 152.

care for others to be what they are. In the cyberpunk aesthetic of an ecology without nature, where only the self-constructed remains, we find a microcosm of the distinctive dilemma that lies at the heart of dwelling upon a wholly enlightened earth, namely, that although humans seemingly encounter nature produced in their image everywhere they look, always and already enframed as a means in service of an endless perpetuation of the self, such a mode of disclosure conceals the fact that "precisely nowhere does man today any longer encounter himself, i.e., his essence."[23] Our global environmental predicament is so uncanny because the successful enframing of the earth corresponds with a complete loss of world, and in effect a concomitant alienation from that which is most intimate.

It is in this particular sense that this book shall operationalize Horkheimer and Adorno's famous thesis that "myth is already enlightenment, and enlightenment reverts to mythology."[24] Despite the prominent status ascribed to instrumental reason in the modern epoch, in whose name the notion of an ensouled nature had to be sacrificed on the altar of progress, modern humans never managed to entirely banish the animistic elements of their primitive past. Hence, if modernity has been premised on the exclusion of such premodern facets, it has, on the other hand, always been haunted by an insistent return of the repressed. From a modern perspective, of course, the repressed would first and foremost appear as an unsettling other — as the irrational forces of that "great enchanted garden,"[25] supposed to have been dispelled once and for all. It is no accident that, from Karl Marx to Jacques Ellul, critics of technology have "shivered [...] before the spectacle of the mechanized proletarian who is subject to the absolute domination of a mechanized capitalism

23 Martin Heidegger, "The Question concerning Technology," in *The Question Concerning Technology and Other Essays,* trans. William Lovitt (New York: Harper & Row, 1977), 27.

24 Horkheimer and Adorno, *Dialectic of Enlightenment,* xviii.

25 Max Weber, *The Sociology of Religion,* trans. Ephraim Fischoff (London: Methuen, 1965), 270.

and a Kafkaesque bureaucracy."[26] Nor that, from gothic horror to cyberpunk, modern humans have been apprehensive of their own synthetic children. In fact, the modern canon is replete with metaphors for the manufactured risks of modernity and the eerie impression of having engineered beings indifferent to the intentions of its artificers, whereby it is precisely this dedication to convert all life to the artificial registers of an anthropogenically stamped form that brings about the return of what modernity has repressed, namely, the impotence of humans to manage and control that which conditions their own existence, without thereby also losing their exceptionalism in the process. Without listing them here, there are countless other examples (fictional and nonfictional) of this scenario in which technological manipulation, through humankind's instrumentalization of nature in an effort to subjugate it, inadvertently threatens the presumed mastery that distinguishes the modern human subject from its other.

As opposed to a confirmation of humankind's narcissistic omnipotence, this is to suggest that the complete artificialization of the earth — insofar as artificial processes of change have now become powerful enough to compete with the global forces of nature — has paradoxically made it "so alien, so complex, so awesome, and so overwhelming that we [...] regress to a degraded state of nondifferentiation from it; this outer reality is psychologically as much a part of us as its poisonous waste products are part of our physical selves."[27] On an unconscious level, "we powerfully identify with what we perceive as omnipotent and immortal technology, as a defense against intolerable feelings of insignificance, of deprivation, of guilt, of fear of death," while giving ourselves "over to secret fantasies of omnipotent destructiveness, in identification with the forces that threaten to destroy the world."[28] It is for this reason that Crutzen, although

---

26 Bruno Latour, *We Have Never Been Modern,* trans. Catherine Porter (Cambridge: Harvard University Press, 1993), 115.

27 Harold Searles, "Unconscious Processes in Relation to the Environmental Crisis," *Psychoanalytic Review* 59, no. 3 (1972): 368.

28 Ibid., 370.

he emphasizes the earth's uncanniness, can nevertheless celebrate that "the long-held barriers between nature and culture are breaking down," and affirm that it is therefore "no longer us against 'Nature,' [but i]nstead, it is we who decide what nature is and what it will be[. … I]n this new era, nature is us."[29] Rather than constituting a way of manipulating forces external to the human subject, there is thus a danger that such a narcissistic injury may instead serve to pave the way for the unrestrained assertion of a will to power precisely by dissolving the boundary between subject and object, leading to "an extension of the power of the will which recalls the 'animistic' conception of the universe that precedes the emergence of the mature ego."[30] With the return of animism in machinic form, we are no longer faced with the coercion of the natural world through the intention of an artificer to subject its forces to mastery.[31] Instead, the continued exploitation of nature may be ontologically sanctioned by locating the will immanently to it. Neither subject nor object, such a force of nature is conceptually converted into unconditional production by and for itself. Importantly, this is not to say that we never really encounter nature in the wake of modernity since our experience is always and already technologically mediated, but, precisely to the contrary, that human artifice is accepted as always and already natural, and that humankind's production is constitutive of nature as such. In other words, nature is taken as in itself nothing but creative production, and so it is precisely by the means of artificially altering its environment that humanity is understood to be acting in accordance with its own nature.

"How antifoundationalism can thus coexist with the passionate ecological revival of a sense of Nature," Jameson writes, "is the essential mystery at the heart of what I take to be a fundamental

---

29 Paul Crutzen and Christian Schwägerl, quoted in Jeremy Baskin, "Paradigm Dressed as Epoch: The Ideology of the Anthropocene," *Environmental Values* 24 (2015): 10.

30 Bukatman, *Terminal Identity*, 210.

31 Angela Melitopoulos and Maurizio Lazzarato, "Machinic Animism," *Deleuze Studies* 6, no. 2 (2012): 240–49.

antinomy of the postmodern."[32] Taking Jameson's provocation seriously, it is the guiding conviction of this book that tackling said "mystery" must be the central task of a questioning of technology adequate to the uncanny experience of our current global environmental predicament. Could it be so that rather than sites of resistance against the instrumentalization of the earth into a standing-reserve, "this becoming organic, or becoming ecological, is no more than the mechanistic-technological triumph of modernity over nature[?]"[33] At the very least, such a question is justified by the suspicion that, as the philosopher Yuk Hui has suggested, "it is no longer a dualism which is the source of danger in our epoch, but rather a non-dualistic totalizing power present in modern technology, which ironically resonates with anti-dualist ideology."[34] Although we mourn the end of nature, it is only all the more important that we do not prematurely grasp for an artificial organicism to re-create some prelapsarian utopia of a synthetic Eden. For if the sublime has migrated from the natural into the artificial, then it is only because the immanentization of human artifice into productive nature signifies the latter's complete technification. As a response to the uncanny affects of dwelling on an artificial earth, the regressive drives of an antihumanist desire to return the human to the natural world—a *regressus ad uterum* on a global level—as a means of escaping alienation, and to form an organic society of symbiotic beings in place of modernity's collection of self-contained bourgeois individuals, can all too comfortably be enrolled in support of the techno-optimistic sentiments of a bright-green ecological modernization. In the anxiety-ridden social reality we find ourselves today, our technology might very well become a conceptual location for intimating the repressed depth of the modern project's failed effort to master nature. But—and this is what

32 Fredric Jameson, *The Seeds of Time* (New York: Columbia University Press, 1996), 46–47.

33 Yuk Hui, "Machine and Ecology," *Angelaki: Journal of the Theoretical Humanities* 25, no. 4 (2020): 59.

34 Ibid., 58.

this book seeks to caution — it can equally well come to serve as the omnipotent object to be fused with and worshipped, a location where sadistic-destructive fantasies of annihilation can run rampant. What we need is thus an engaged questioning of technology — tied to social practice and theory — that seeks to clarify the corruptibility of the synthetic merge between natural geomorphology and human artifice to regress into the latter as opposed to progressively contribute to the former. What follows is a genealogical attempt at making such a clarification.

From Hutton to Lovelock and Back Again

To accomplish the task now set before us, chapter 1, "Toward a Terrestrial Turn?," introduces the concept of planetary technicity by investigating the methodological transformations that set the scene for a heightened awareness of global environmental change in the 1980s, and out of which the now widely debated Anthropocene and the variously associated ontological claims about the hybrid nature of our artificial earth have subsequently taken shape. The study of the history, sociology, and philosophy of global change research — particularly meteorology and atmospheric science — has exploded during the last twenty years, but far less attention has been paid to the hermeneutic question of how transdisciplinary efforts, such as earth system science, have disclosed humanity's relationship to its planetary abode. Yet, the application of systems theoretical tools to conceive of the planet as an interacting whole has as of late come to play a remarkably influential role — scientifically, culturally, and politically. It has served to prove the capability of treating complex systems with computer simulation — a breakthrough for the earth sciences and beyond — and has been invoked as a source of scientific confidence and authority. Moreover, it has become visible and famous in the public sphere, has helped to spawn a renewed interest in and debate about the growing effect of humankind upon the biosphere, and has led to calls for a novel political paradigm of earth system governance. In 2001, the Amsterdam Declaration on Global Change declared that "the Earth System behaves as a single, self-regulating system comprised of physical, chemical, biological and human components," and that, because of its dynamic behavior, "global change cannot be understood in terms of a simple cause-effect paradigm."[35] Cementing the position taken in 2001, a second conference on global change was held in London in March 2012, again emphasizing that "the Earth system is a complex, interconnected system that includes the global economy and society, which are themselves highly interconnected and interdependent."[36] Consequently, biological and technological processes have been conceptualized as integral parts of the earth system rather than mere passive recipients of changes in the geospheres. This includes alterations in and by the nitrogen and carbon cycle, atmospheric composition, and marine food chains, but also technical infrastructures, such as transport, communication, and urbanization. To this extent, it has been argued that "the Earth System includes humans, our societies, and our activities" and that "humans are not an outside force perturbing an otherwise natural system but rather an integral and interacting part of the Earth System itself."[37] Conceptually integrating technology into the larger terrestrial environment has thusly been identified as being decisive for properly addressing environmental challenges on a global scale.

In an effort to excavate certain moments where contemporary ideas about the hybrid nature of our artificial earth find historical resonance, chapters 2 through 4 trace the genealogy of planetary technicity all the way up to the birth of earth system science in the 1980s. Beginning with the scientific formaliza-

35 Jan Pronk, "The Amsterdam Declaration on Global Change," in *Challenges of a Changing Earth. Global Change — The IGBP Series*, eds. Will Steffen et al. (Berlin: Springer, 2002), 207.

36 Lidia Brito and Mark Stafford Smith, *State of the Planet Declaration — Planet under Pressure: New Knowledge towards Solutions Conference, London, 26–29th of March 2012* (London: Diversitas, 2012), 6.

37 Will Steffen, Paul J. Crutzen, and John R. McNeill, "The Anthropocene: Are Humans Now Overwhelming the Great Forces of Nature?," *Ambio* 36, no. 8 (2007): 615.

tion of the discipline of geology out of late eighteenth-century natural philosophy, chapter 2, "Deep Time of the Heat Engine," focuses on the introduction of the concept of self-organization into geology through the geotheory of James Hutton, paying particular attention to his ambiguous depiction of the earth as simultaneously a machine and an organism. By putting Hutton's geotheory in the context of the Romantic portrayal of human artifice as an expression of seemingly natural processes of deterioration and regeneration, chapter 2 examines how metaphors for technology shifted away from the dead mechanism characteristic of clockwork to the kind of living feedback that would later come to be formalized in thermodynamics, but that, in the late eighteenth century, was already familiar to savants such as Hutton in terms of the organic body. In general, organicism has been regarded as inherently at odds with instrumentalism: the latter, by all accounts, reduces the natural world to its use for human purposes, and the former operates on a desire to reconcile nature with the human by stressing a much deeper interconnection between both. But even though they condemned the narrow-minded instrumentalism of industrial modernity, the Romantics did not abandon the commitment to technology per se; rather, what they rejected was the insufficiency of the instrumentalist interpretation of that commitment, proposing in its stead a different perspective from which to understand the relationship between nature and artifice. Because of this proposed change in perspective, nature was no longer something that could be judged from a particular point of view. Rather, nature could only be comprehended as a complex whole, which, moreover, meant that human artifice, as part of nature, had to be understood as participating in the universal history of the earth itself. Chapter 2 cautions that along with such a change in perspective, however, any sense of a limit—such as a horizon of understanding belonging to human history—thereby disappears into the abyss of geological time, and the subject suddenly vanishes from the center of the global environmental drama. Ironically so, since the purported novelty of the globalization of technology is precisely the manner in which it highlights the anthropogenic dimension of global environmental change, and thus the deep time consequences of human action.

But the meeting between Romanticism and the burgeoning geological sciences around the turn of the eighteenth century is far from a lone instance in modern intellectual history when it comes to reconceiving human artifice from that of an external imposition upon the earth to something much more akin to an artful disclosure of its inner potential. In its wake, several other intellectual heavyweights continued the project of further unearthing the significance of the role of technology in planetary evolution. In fact, the human/nature coupling was strongly emphasized and promoted by two scholars at the beginning of the twentieth century, one of them the Russian mineralogist and geochemist Vladimir Vernadsky, who published a series of lectures on the subject, titled *The Biosphere* (1926). In these lectures, Vernadsky developed an integrative and functionalist definition of the planet to comprise both living beings and the nonorganic matter sustaining them—including, he argued, technology, with the help of which humankind had become such a crucial component of the earth that it could no longer be ignored as a geological force.

Proceeding from Hutton's ambivalent oscillation between machine and organism, chapter 3, "Dissolving Technology, Planetary Metamorphosis," examines how the topological function of the sphere as an operational interface between biotic and abiotic matter came to influence the understanding of technology by bringing not only organisms but also artifacts into natural evolution. Along with the study of global biogeochemical cycles and the concomitant recognition of humanity's growing effect on the biosphere, speculations on the nature of technology surfaced in the intellectual circles of Paris during the interwar period—certainly through the work of Vernadsky, but also in the work of the Jesuit paleontologist Pierre Teilhard de Chardin. Focusing on Vernadsky's holistic and integrative approach to the study of process on the level of biosphere, chapter 3 observes how this approach laid the foundation not only for the study of anthropogenic environmental change but also for an understanding of human artifice as a functional extension of the singular process of the earth's self-organization. Moreover, Vernadsky's development of a biogeochemical approach to the study of the earth is juxtaposed and analyzed in relation to Teilhard de Chardin's speculative anthropology by tracing their theoretical indebtedness to the Bergsonian philosopher Édouard Le Roy's orthogenic view of terrestrial evolution. Together, their writings spawned a heady mix of a multiplicity of overlapping perspectives — borrowing from scientific, cultural, and explicitly religious genres — through which the unfolding of this understanding of humankind's being on the earth entailed the transformation of age-old oppositions and a number of boundary breakdowns. However, chapter 3 concludes by cautioning that even though the flattening of the modern philosophical division between nature and artifice portrays itself as an ontological corrective to that insufficiently materialist dualism underlying mechanic philosophy, it ironically remains an *idealism* in the most fundamental sense of that word. Because to fill the inorganic inwardly with spirit is, as the philosopher Louis Althusser famously warned, to smuggle idealism into materialism, upon which one may then justify class relations, bourgeois politics, and the apparatuses of capital through reification. Only in accordance with such an organicist ontology could instrumentalism be set free from its utilitarian constraints of a mere means to become mythologized into an end in itself.

Still, Vernadsky's teachings remained relatively obscure in the West until G.E. Hutchinson popularized them in the latter part of the twentieth century, at around the same time that Vernadsky was called "the father of modern biogeochemistry" by the British atmospheric chemist James Lovelock, who, in his own right, went on to propose that feedback in the climate system was intricately connected to the homeostasis of basic geophysical processes. From the development of this feedback-based, integrative science, which Lovelock himself, following Hutton's metaphor of the body, called "geophysiology," sprang a number of interesting reflections on the essence of technology. As a product of their collaborative work in the 1970s, the

"Gaia hypothesis" was advanced by Lovelock together with the American microbiologist Lynn Margulis as a means to provide an ontological basis for integrating all components of the earth system, thereby reviving the Huttonian idea of the planet as a self-organizing entity, but now under the auspices of the cybernetic notion of the thermostat. The earliest versions of the Gaia hypothesis contained phrases such as "by and for the biosphere," thereby implying the sense of a joint purposefulness on the part of life in general to artificially produce the global environment in ways that suited its continued existence, thus facing the controversial question of teleology in nature head on. The genealogical investigation is thereby closed out in chapter 4, "Mythology in the Space Age," by examining how the figure of technological life reappeared in cybernetic discourse during the Cold War with the associated propagation of systems science for the sake of global military surveillance and control. Reengineering the earth's future along the lines of positive and negative feedback loops, Lovelock and Margulis shamelessly reintroduced natural teleology at the heart of their twentieth-century resurrection of the geotheoretical tradition, in effect reimagining the ontological status of the artifact, away from that of an anthropological instrument and instead toward constituting the primary milieu of the organism. If it has been far too common in contemporary philosophy of technology, especially in its critique of instrumentalism, to frame the concern with the globalization of technology in terms of the dominance of exploitative-egoistic Cartesianism over neopagan Spinozism, chapter 4 argues that the Gaia hypothesis of Lovelock and Margulis constitutes an exceptional case of a boundary object, curiously enrolled by both New Age spiritualists and Promethean ecomodernists. Put differently, there is a surprisingly small step from Gaia as a metaphor for vulnerability and community to one that describes the technological realization of a nature yet to come — to be actualized poietically by the biota, as Lovelock and Margulis imagined it. Indeed, from the Gaian point of view, ontic beings, including humans, exist as but elements in more-than-human configurations of energy transformation, whose goal, in what can best be described as a kind of Nietzschean ecology of self-overcoming, is nothing but the intensification of the vital impulse to self-organize in increasingly complex patterns.

The book's conclusion, "The Will to Terraformation," brings the insights from the genealogical examination of planetary technicity to bear on a critique of the present. With reference to the preceding chapters, it argues that it is not because global technology is gradually becoming more seamless and more indistinguishable from nature's forces that the barrier between what is considered "natural" vis-à-vis "artificial" has seemingly collapsed, but rather that it is because the collapse of the barrier between what is considered "natural" vis-à-vis "artificial" has a priori come to dictate our horizon of experience that global technology is seemingly becoming more seamless and more indistinguishable from nature's forces. One of the consequences of the genealogy presented in this book is thus to nuance the etiology of the Anthropocene provided by its proponents, who, although they generally agree that it was the industrial revolution and its consequences that inaugurated this new epoch of natural history, nonetheless hold that earth system science is responsible for raising humanity's self-awareness to this "scientific fact." Although the critique presented in the conclusion to this book is not meant to dispute the crucial role played by earth system scientists in making global environmental change into a matter of concern, nor the plethora of risks associated with humankind's ability to alter the conditions for life, it nevertheless makes use of this genealogy to stress that empirically verifiable patterns of anthropogenic environmental change, no matter how detailed and well documented, cannot elucidate the ontological dimension to the Anthropocene condition. Last, it cautions that there is an ever-present corruptibility to the synthetic merge between natural geomorphology and human artifice that consists in the reinstatement of an intrinsic teleology in which technology takes on the central role as a transcendental signified, and that anchors and secures the meaning of being—albeit nihilistically so—in an unrestrained instrumentalism. In place of the transcendent artificer, it is argued, we instead get a self-developing "will to terraformation" internal to nature itself. This is an intellectual lineage that runs through the work of Hutton, Vernadsky, and Lovelock, a mythos upon which planetary technicity, so central to the earth system paradigm, operates.

§

There is a rich lineage of reflections on the essence of technology that runs through the history of earth science, which has revolved around efforts to widen technology beyond its reduction into the supplementary status of an instrument, instead emphasizing its character as a global force on par with the rest of the earth's geospheres. As we shall see, it was largely thanks to the geotheory of Hutton that the groundwork for the reinterpretation of human artifice as a part of the self-organizing capacity of the earth had already been laid in the late eighteenth century. This foundation was then built upon in the early twentieth century when Vernadsky, together with Teilhard de Chardin, worried about the relationship between organic and inorganic processes for the evolution of life on earth and proposed a vision of human artifice not just as an imitation of nature but as an elementary manifestation of the integrative function and evolution of the terrestrial environment. Finally, in the second half of the twentieth century, the coevolution of organic and inorganic processes was further developed upon, this time in cybernetic terminology, and postulated by Lovelock and Margulis as the foundation for planetary homeostasis.

Proceeding from a genealogical point of departure, the ambition of this book is to historically examine how the study of the earth led to reflections on the essence of technology, and how these reflections, in turn, altered beliefs in and caused changes to the accepted explanations of the structure and composition of the planet and humanity's relationship to it. Accordingly, this book seeks to supply a richly recollected and historically reflective dimension to the consolidation of the global environment into the systems-theoretical paradigm of earth system science, and to the associated Anthropocene discourse on humanity's relationship to the earth. It seeks, in other words, to usher it into its phase of critical self-consciousness. Surveying and thematizing the concern of defining, describing, and delineating the role of humankind as both observer and participant in the geological economy, this book delves into the conceptual realm that constitutes the self-reflexive dimension of the discipline of earth science. As a consequence, it has a dual aspect: it seeks not only to reconstruct a catalogue of explicit meditations within the earth sciences upon technology as a global phenomenon, but also to provide an in-depth and long-durational genealogy of the discursive conditions underwriting the synthesis of nature and artifice within geophilosophical registers.

What will be attempted herein is thus a study of what characterizes our present concerns about technology in the face of global environmental change by exploring an intellectual legacy that has largely been neglected in conventional historical and philosophical treatises on technology. Such an interdisciplinary cross-pollination between philosophy and history into the framework-explicating impetus of a critical genealogy concerns itself with lineages of a conceptual nature that then become embedded in discursive practices and vocabularies, such that one can wield them without having a detailed understanding of where they came from. But this book is necessarily interdisciplinary also in an additional sense, because the fact of the topic — the disclosure of technology as a global phenomenon — evidently emerges as a confluence of multiple technical lexicons across various domains. Tracing its provenance from the natural-theological concept of a self-organizing earth that fueled the Huttonian systematization of geology all the way to the global environmental concerns of the twentieth century, and thus across multiform encounters between philosophy and earth science, this book orients itself around the concept of planetary technicity as a guiding thread to rediscover overlooked pathways in modern thought. It suggests that, far from being an abstract concern unrelated to advances in the earth sciences, the question concerning the essence of technology dramatizes fundamental philosophical problems of subjectivity, freedom, and the transcendental that remain central to the modern attempt to reconcile human experience with the scientific discoveries about the natural history of our planet. If we are caught between a rock and a hard place when trying to make sense of the essence of technology today, then a careful consideration of the history of its ontology could contribute to a thematic outlook of enduring relevance.

1

Toward a Terrestrial Turn?

Over the last two decades, it has become a generally accepted claim among earth scientists that in the twentieth century a new component of the earth established itself with the emergence of technology as a multiscalar system, comparable in the reach and range of its effects to the planet's geospheres.[1] Nowadays, only the most hostile environments — ice-covered land and seas, deserts, and some of the densest areas of rainforest remote enough from human populations — can still be regarded as near-pristine. In fact, the rate of human-induced change to land, marine, and atmospheric environments, along with the equally worrying fact that these changes have now become discernable on the global scale, have prompted the proposal within the academic community that we stratigraphically adopt the term "Anthropocene" to describe the epoch within which we currently live.[2] This is more than a lighthearted neologism to describe an accelerated rate of anthropogenic environmental change. Stratigraphers are seriously engaged in international

1 The geophysicist Peter K. Haff has proposed that we call this phenomenon "the technosphere"; see Haff, "Technology as a Geological Phenomenon: Implications for Human Well-Being," *Geological Society Special Publication* 395, no. 1 (2013): 301–9.

2 Paul J. Crutzen and Eugene F. Stoermer, "The 'Anthropocene,'" *IGBP Global Change Newsletter* 41 (2000): 17–18.

discussions about the merit and feasibility of defining a subdivision of the Geologic Time Scale in accordance with human perturbation of the global environment, and thus the designation of the Anthropocene as a formal time unit.[3] The suggested Anthropocene epoch denotes an unofficial interval of geologic time that would constitute the third worldwide division of the Quaternary period, after the Pleistocene and Holocene, to mark the beginning of when human artifice started to have a significant effect on the planet's geology and ecosystems, and to such a degree as to match or even surpass the earth's natural forces of change. In one of the first papers to expand on the concept, the atmospheric chemist Paul Crutzen proposed that the Anthropocene could be said to have begun sometime in the latter half of the eighteenth century, as analyses of air trapped in polar ice suggests a rise in global concentrations of carbon dioxide and methane during this time, which, befittingly, coincides with James Watt's design of the steam engine in 1784. Since then, fossil fuel burning and the intensification of agriculture have caused an increase in the concentration of carbon dioxide by more than 30 percent, and methane by more than 100 percent, the global methane-producing cattle population has risen to 1.4 billion, and tropical forests have disappeared at an alarmingly fast rate, releasing carbon dioxide and further diminishing carbon sequestration.[4] The underlying fear is that rapidly progressing and unabated environmental change will lead to a global crisis for humankind, because it means that the relatively stable climatic era since the last ice age, in which human civilization has thus far developed, will come to an end. During the past 2,000 years, fluctuations in the mean global temperature have amounted to less than one degree Celsius, which, according to the chemist and climate scientist Will Steffen and his colleagues, ought to worry us, since neither our agriculture and forestry

3 Jan A. Zalasiewicz et al., "The New World of the Anthropocene," *Environmental Science and Technology* 44, no. 7 (2010): 2228–31.

4 Paul J. Crutzen, "Geology of Mankind," *Nature* 415, no. 6867 (2002): 23.

nor our social infrastructures have been designed to withstand a rapid and significant change of several degrees.[5]

But the apprehension about environmental change captured by this new geological epoch is not limited to a global increase in the release of greenhouse gases and a resulting concentration of carbon dioxide in the atmosphere. Rather, the extent of environmental change caused by technological processes has altogether reached another dimension since the period of industrialization. Although a rise in the average temperature of the earth's climate system plays a significant role in altering the conditions for life on the planet, the magnitude, variety, and longevity of technologically induced change mean that the activities of humankind reach far beyond the borders of any single geosphere. In fact, humanity has had a dramatic effect on the entire global environment, having reengineered around half of the world's land surface. Forests, savannahs, and grasslands are being cleared to make way for agriculture at an increasing speed, and along with a growing global population, increased bioenergy production and biomass use, and an expanding infrastructure, this has amounted to growing pressures on land use.[6] These global change patterns pertain not only to land, because, similarly, humans now use more than 40 percent of the renewable, accessible water resources.[7] Collectively, various anthropogenic global material and energy fluxes now by far exceed any natural flows, and as a consequence developments in many of the vital environmental dimensions are reaching a crisis stage. Water resources, soils, forests, and oceans have been overexploited or are being destroyed, biodiversity is undergo-

5 German Advisory Council on Global Change (WBGU), *World in Transition: A Social Contract for Sustainability. Flagship Report 2011* (Berlin: WBGU, 2011), 33, and Will Steffen et al. (eds.), *Global Change and the Earth System: A Planet under Pressure—IGBP Global Change Series* (Berlin: Springer, 2005), 209–13.

6 Detlef P. van Vuuren, *Growing within Limits: A Report to the Global Assembly 2009 of the Club of Rome* (Bilthoven: PBL Netherlands Environmental Assessment Agency, 2009).

7 Millennium Ecosystem Assessment (MEA), *Ecosystems and Human Well-Being: Current State and Trends* (Washington, DC: Island Press, 2005), 1:167.

ing a drastic reduction, and important biogeochemical flow patterns have been radically altered by humankind.[8] In addition to large-scale changes to land use, human civilization has already caused widespread species extinctions,[9] species invasions,[10] and changes in the local and global cycles of carbon, nitrogen, phosphorus, and other elements.[11] Since the industrial revolution, the trajectory of the human population has also been one of steady and rapid growth, skyrocketing from just under 1 billion to almost 8 billion. Similarly, yearly energy consumption per capita, although it does not amount to more than around 6.438 kWh per capita in India, is almost ten times as high in some countries in the Global North, such as Sweden. An important driving factor behind this expansion has undoubtedly been the burning of fossil fuels, having made possible both intensive agriculture and a massive increase in material flow, which, in industrial societies, amounts to about ten to thirty tons per person every year.[12] More than half of all accessible fresh water and about 30 to 50 percent of the planet's land surface is now in use by humans, energy use has grown sixteenfold during the twentieth century, more nitrogen fertilizer is applied in agriculture than is fixed in all terrestrial ecosystems, and the anthropogenic emissions of sulfur and nitric oxide overrides naturally occur-

8 WBGU, *World in Transition*, 31.

9 Rodolfo Dirzo et al., "Defaunation in the Anthropocene," *Science* 345, no. 6195 (2014): 401–6, and Stuart L. Pimm et al., "The Biodiversity of Species and Their Rates of Extinction, Distribution, and Protection," *Science* 344, no. 6187 (2014): 1–10.

10 Peter M. Vitousek et al., "Introduced Species: A Significant Component of Human-Caused Global Change," *New Zealand Journal of Ecology* 21 (1997): 1–16, and Anthony Ricciardi, "Are Modern Biological Invasions an Unprecedented Form of Global Change?," *Conservation Biology* 21, no. 2 (2007): 329–36.

11 Peter M. Vitousek et al., "Human Domination of Earth's Ecosystems," *Science* 277, no. 5325 (1997): 494–99; Paul Falkowski et al., "The Global Carbon Cycle: A Test of Our Knowledge of the Earth as a System," *Science* 290, no. 5490 (2000): 291–96; and James N. Galloway et al., "Nitrogen Cycles: Past, Present, and Future," *Biogeochemistry* 70, no. 2 (2004): 153–226.

12 WBGU, *World in Transition*, 34–35.

ring ones.[13] It is not for nothing that the Board of the Millennium Ecosystem Assessment in 2005 came to the conclusion that "human activity is putting such strain on the natural functions of the earth that the ability of the planet's ecosystems to sustain future generations can no longer be taken for granted."[14] From the perspective of the transdisciplinary endeavor established to study and address the planetary reach of human activity, "earth system science," it is no longer possible to describe or predict ecosystem interactions without considering the role of technology — not only as an external variable, but also as *internal to* and *constitutive of* many of the processes that we have hitherto called "natural."[15]

In light of this radical change to the reach and range of technical alteration, the coevolution of humankind and the biosphere has become one of the principal questions of our age. As we find that humanity has altered the planet at just about every scale we are capable of measuring, and as we find ourselves having entered the sixth great mass extinction in earth's natural history,[16] an event that may even claim our own species as one of its many victims, the question concerning the essence of technology, in its power to not only imitate but in many ways even surpass the forces of nature, has become critical for the discussion about global environmental change. We need only to consider the global environmental challenges that we currently face — an accelerating loss of biodiversity, degradation of land and fresh water, rapidly changing precipitation patterns, increasingly frequent extreme weather events, declining permafrost, to name a few — for it to become evident how these are

13 Crutzen, "Geology of Mankind," 23.

14 Millennium Ecosystem Assessment (MEA), *Living Beyond Our Means: Natural Assets and Human Well-Being: Statement from the Board* (New York: MEA, 2005), 5.

15 Erle C. Ellis and Peter K. Haff, "Earth Science in the Anthropocene: New Epoch, New Paradigm, New Responsibilities," *EOS Transactions* 90 (2009): 473, and Anthony D. Barnosky et al., "Approaching a State Shift in Earth's Biosphere," *Nature* 486, no. 7401 (2012): 52–58.

16 Anthony D. Barnosky et al., "Has the Earth's Sixth Mass Extinction Already Arrived?," *Nature* 471, no. 7336 (2011): 51–57.

increasingly experienced as indications of an osmosis between nature and artifice: between the artificial products and processes of humans, on the one hand, and the natural products and processes of the earth, on the other, gradually suffusing into each other to create an amorphous and indeterminate hybrid of the two. As industrialized humankind has literally become a "force" to be reckoned with on the geological scale of deep time, an agent exerting a domineering influence on par with natural processes of change, issues of global environmental change appear to us both as consequences of our activity qua geophysical force and as an urgent and inescapable demand to take responsibility for the faltering sustainability of our terrestrial life-support system. It is this puzzling condition, the seeming fusion between nature and artifice in the discourse of the Anthropocene, that is the central topic of this chapter.

Our Anthropocenic Milieu, or, a Cybernetic Ecology

Central to the Anthropocene proposition is the claim that we have left the relatively benign period that was the Holocene — a period during which human civilization developed and thrived within a relatively stable earth system — and have entered a significantly more unpredictable epoch, one in which technology has become so ubiquitous — globally — as to alter the very planetary life-support systems upon which humanity has hitherto not only depended but has taken for granted as fundamental and unalterable properties of its "natural" environment. It has been suggested, in fact, that we have already overstepped no fewer than three of nine interlinked global thresholds that, when exceeded, may lead to intensifying feedback loops and runaway scenarios that could leave us with irreversible alterations to the dynamics of the earth system. In order to press home the point of precisely how artificial our earth has become, the concept of "planetary boundaries" has since 2009 been developed by a group of earth system scientists led by Johan Rockström. These boundaries are defined as quantitative damage thresholds whose transgression could potentially generate dangerous ripple effects throughout the entire earth system. Recently, for instance, the German Advisory Council on Global Change drew from the same insights to promote the adoption of "planetary guard rails," which they describe as "quantitatively definable damage thresholds whose transgression either today or in future would have such intolerable consequences that even large-scale benefits in other areas could not compensate these."[17] Although these boundaries are not intended to demarcate exactly defined limits within which there is hardly any risk at all, or beyond which we can immediately expect serious damage and disaster, they nevertheless serve to emphasize our need to avoid "tipping points," that is, irreversible changes to the structure of the earth, such as the melting of Greenland's ice sheet, the collapse of tropical coral reefs, or other nonlinear processes whereby changes to global systems risk becoming self-propelling.[18]

Admittedly, the social and political implications of the Anthropocene are in many ways a revival of the Hobbesian preoccupation with questions of tolerance and the accommodation of a plurality of differing and competing beliefs in the face of civil unrest. In any case, the legislation of limits is absolutely central. From Thomas Malthus's theory on population growth, through Garrett Hardin's concern for our commons[19] and the carrying capacity of ecosystem services, all the way up to modern analytical concepts, such as planetary boundaries, there is a long tradition within Western thinking concerned with the effort of restricting humanity's consumption and modification of nature within constraints that guarantee its future maintenance, regardless of whether such constraints are defined through "limits," "thresholds" or "loads," or if they are global or local in scale. But whereas the Hobbesian tradition takes the doctrine of original sin as its starting point and proceeds

---

17 WBGU, *World in Transition*, 32.

18 Johan Rockström et al., "Planetary Boundaries: Exploring the Safe Operating Space for Humanity," *Ecology and Society* 14, no. 2 (2009): art. 32.

19 "The commons," as in a natural resource available to all members of a society, notoriously difficult to privatize. See, for instance, Garrett Hardin, "The Tragedy of the Commons," *Science* 162, no. 3859 (1968): 1243–48.

from a divinely given distinction between nature and artifice in the wake of humanity's fallenness, the idea of an Anthropocene, rather, represents the toppling of the primordial state of nature into an already artificial earth. In this day and age, we have passed out of our natural environment and instead live in a fully technological setting, which means that the very idea of nature has become questionable. Rather, the global technological infrastructures of the new millennium now constitute our natural milieu. Instead of denoting a separate sphere of human life, technology has become a properly global system of which we as humans are but a part.

Drawing upon the notion of the dynamic interaction between geospheres, several scholars, following the geophysicist Peter Haff, have proposed that the spatial manifestation of the Anthropocene consists in the emergence of a new sphere — what they call "the technosphere." Stemming from the entanglement between natural and artificial environments, the technosphere forms a new and highly dynamic component of the earth, giving rise to cybernetic ecologies that drastically change the metabolism of our planet — amorphous in its Gestalt, yet powerful enough to alter the entire history of the planet and its conditions for life.[20] Seen from Haff's geophysical perspective, the

20 Katrin Klingan and Christoph Rosol, eds., *Technosphäre: 100 Years of New Library* (Berlin: Haus der Kulturen der Welt, 2019). See also Haff, "Technology as a Geological Phenomenon." It is worth mentioning, as a number of scholars have already noted, that Gilles Deleuze and Félix Guattari already presaged a similar notion — what they called "the mechanosphere" — in their 1980 collaborative book *A Thousand Plateaus*. See, for instance, Arun Saldanha, "Mechanosphere: Man, Earth, Capital," in *Deleuze and the Non/Human*, eds. Jon Roffe and Hannah Stark (Berlin: Springer, 2015), 197–216; Hunter Dukes, "Assembling the Mechanosphere: Monod, Althusser, Deleuze, and Guattari," *Deleuze Studies* 10, no. 4 (2016): 514–30; and Arun Saldanha and Hannah Stark, "A New Earth: Deleuze and Guattari in the Anthropocene," *Deleuze Studies* 10, no. 4 (2016): 427–39. The mechanosphere asks us to take seriously nature and artifice as two branches of the same (in Deleuze and Guattari's terms) "abstract machine." In this sense, the human circulatory system, the circulation of capital, and the global carbon cycle are not ontologically distinct processes but mutually constitutive assemblages; see Gilles Deleuze and Félix Guattari,

emergence of the technosphere represents a novel evolutionary event in the earth's history that has historically reshaped and will continue to reshape the biosphere for the foreseeable future,[21] in other words, a long-term dynamics of anthropogenic ecological patterns and processes in landscapes, ecosystems, and biogeography, including the transformation of biomes into "anthromes."[22] In addition to its natural processes, our planet now includes an amorphous fabric of artifice, which consists of all the technical configurations that have become integral parts of our daily lives—everything from industry, housing, transportation, information and communication systems, farming, and mining, to landfills and spoil heaps. In an article published in the *Anthropocene Review,* Jan Zalasiewicz—the chair of the Anthropocene Working Group of the International Commission on Stratigraphy, tasked with considering the Anthropocene as a potential addition to the Geologic Time Scale—argues,

---

A Thousand Plateaus: Capitalism and Schizophrenia, trans. Brian Massumi (Minneapolis: University of Minnesota, 1987), 71. The biosphere has created the conditions for the emergence of the mechanosphere, but those machinic assemblages have the capacity to turn back and modify their own conditions. Thus, the networks of machines ("real" and "abstract") that now cover the earth unavoidably intervene in its geophysical systems, rendering obsolete any rigid demarcation between "primary" and "secondary," "original" and "prosthesis." See Deleuze and Guattari, *A Thousand Plateaus,* 69. As we shall see in chapter 3, however, both the technosphere and the mechanosphere bear a striking resemblance to the concept of "the noösphere," most thoroughly developed by Vladimir Vernadsky and Pierre Teilhard de Chardin during the mid-twentieth century. Although Deleuze and Guattari are keen to distinguish the mechanosphere from the noösphere since, as they argue, the latter "introduces a kind of cosmic or even spiritual evolution from one [geosphere] to the other, as if they were arranged in stages and ascended degrees of perfection," in genealogically tracing the roots of such an idea of a global proliferation of technology, I shall instead seek to underscore how their assertion of organic and inorganic entwinement was present in the concept of the noösphere already from its genesis. See Deleuze and Guattari, *A Thousand Plateaus,* 69, 79.

21 Haff, "Technology as a Geological Phenomenon," 302.

22 Erle C. Ellis and Navin Ramankutty, "Putting People in the Map: Anthropogenic Biomes of the World," *Frontiers in Ecology and the Environment* 6, no. 8 (2008): 439–47, and Erle C. Ellis, "Ecology in an Anthropogenic Biosphere," *Ecological Monographs* 85, no. 3 (2015): 321.

together with a number of his colleagues, that the totality of the earth's human-made structures is absolutely astounding in its scale, with some estimations suggesting a mass of more than 30 trillion tons, which, if evenly distributed over its surface, would amount to no less than 50 kilos per square meter.[23] Following in the footsteps of the nineteenth-century geologist Antonio Stoppani, who already in 1873 imagined the stratification of layers of technological rubbish,[24] Zalasiewicz speculates that many of our human-made structures, if entombed in strata, may be preserved into the distant geological future as "technofossils," and help future generations to characterize and date the onset of the Anthropocene.[25]

But there is more to this bulk of artifice than just its mass. In order to properly grasp the idea of the technosphere, we must not get hung up on the question of quantity. Rather, what can be said to distinguish the idea is precisely the dynamic properties it ascribes to technology. The geophysical significance of technology consists in the incessant manufacture of a vast range of artificial objects, which themselves have given rise to entirely novel flows of energy, matter, and information: from simple goods and tools to facilities for energy capture, modes of transportation, media for the storage and sharing of information, and the most sophisticated apparatuses for manipulating nature on a microscopic scale. According to proponents of the Anthropocene, we should thus consider our institutions and forms of organizing social life as constituting a crucial part of this phenomenon; that is, we have become so bound up in its reproduction that we need to maintain its procedure in order to survive, and much like any other complex system, the technosphere comes with its

23 Jan A. Zalasiewicz et al. "Scale and Diversity of the Physical Technosphere: A Geological Perspective," *The Anthropocene Review* 4, no. 1 (2017): 11.
24 Antonio Stoppani, "First Period of the Anthropozoic Era," trans. and eds. Valeria Federeighi and Etienne Turpin, in *Making the Geologic Now: Responses to the Material Conditions of Contemporary Life*, eds. Elizabeth Ellsworth and Jamie Kruse (Brooklyn: punctum books, 2013), 38.
25 Zalasiewicz, "Scale and Diversity of the Physical Technosphere," 16–17.

own internal dynamic of energy flows.[26] Hence, the term also encompasses the enclosure of human populations, forests, cities, seas, and other traditionally nontechnical entities within systems of technical management and productivity. At their current scale, this means that such artificial systems are major, new phenomena in the history of our planet — ones that are evolving extraordinarily rapidly. These "novel ecosystems" — or, again, "anthromes" — are systems that are not naturally occurring, but rather engineered by humans or created as the result of human actions. Anthromes thus signal the mobilization and hybridization of energy, material, and environments into a planetary system on par with other geospheres, emphasizing the leading role of the technological within the global structure of the earth. Elements of these systems incorporate everything from rich soils created by domesticated livestock to nuclear waste, and they include invasive species transported by boat or viruses transported by plane, and the environmental imprint of entities, such as steelworks and industrial factories. In short, these are processes at the intersection between nature and artifice, merging the biological with the technological.[27] They suggest that technological systems in many ways possess the same self-organizing capacities previously ascribed to ecological systems, in that there emerges, within these systems, regulatory capacities on spatial and temporal scales beyond human control.

Because of this, the technosphere dominates through a kind of apparatus of political and social control whereby power resides within its technological matrix,[28] making us all slaves to a system that is at once our life support and a global existential force of environmental alteration. Because of its systemic character, Haff echoes Jacques Ellul in rejecting the notion that we are now ruled by a technocracy — since technocrats merely

26 Peter K. Haff, "Being Human in the Anthropocene," *The Anthropocene Review* 4, no. 2 (2017): 103–9.
27 Ellis and Ramankutty, "Putting People in the Map."
28 Erich Hörl, "Erich Hörl: A *continent.* Inter-view," *continent.* 5, no. 2 (2016): 27, https://continentcontinent.cc/archives/issues/issue-5-2-2016/erich-hoerl.

watch over their limited realms of expertise, it is the technosphere itself that ultimately rules. Subverting the critique of instrumental reason, Haff prefers instead to call us "captives," arguing that "in the technological world of the Anthropocene, most people are subject to the rules of—are essentially captives of—large systems that they cannot control—a corporation, a state, transportation networks, the technosphere as a whole."[29] In short, the technosphere assembles and organizes our lifeworld not according to some higher set of values, but merely for the sake of its own self-perpetuation. Autonomy, here, is key, and reflects the necessities of a system too large for human understanding, but one that nevertheless subsists on its own, without any single individual in control. We intuitively seem to be in control of technology, but this holds only locally, according to Haff. Thanks to technological development—with the instruments and computing devices necessary for measuring and calculating environmental changes across ever-larger timescales—we are becoming increasingly aware that we are ourselves nothing more than parts of our technical systems, as if we are merely moved along by their global metabolism. Of course, humanity may still command authority on a local level, but at the global scale, the system runs itself, and it does so without primary regard for human concern. In this way, the "technosphere resembles the biosphere—complex and leaderless."[30] Unquestionably, humans still make up an important part of the technosphere, but only as another resource to be extracted:

> The technosphere is a system for which humans are essential but, nonetheless, subordinate parts. As shorthand we can say that the technosphere is autonomous. This does not mean that humans cannot influence its behavior, but that the technosphere will tend to resist attempts to compromise its function[. …] It is a global system whose operation underpins

29 Peter K. Haff, "Humans and Technology in the Anthropocene: Six Rules," *The Anthropocene Review* 1, no. 2 (2014): 129.

30 Ibid., 132.

> the Anthropocene and therefore merits special attention in our attempts to understand the role of humans in a nascent geologic epoch.[31]

The metabolism of the technosphere requires massive amounts of resources, which is precisely what makes it a core component of the Anthropocene condition. But if not fed, it will shut down systems that humans now depend on to survive. For the current world population, Haff insists, "is deeply dependent on the existence of the technosphere. Without the support structure and the services provided by technology"[32] we would face a major demographic collapse. In other words, humankind has no choice but to continue that which comes natural to it—namely, altering its milieu—and such a process should not be understood as an external imposition by human artifice upon the terrestrial environment, but as the same kind of terraforming that all organisms are engaged in. The technosphere is humanity's natural environment—it constitutes the preconditions for our modern existence—and so adapting it, and adapting to it, is no less an environmentalist task than adapting, and adapting to, say, the biosphere or the atmosphere. In other words, it has become the natural environment within which human existence, as we know it, is possible. Among the most crucial resources for its perpetuation, then, we find fossil fuels. But equally important, from Haff's point of view, is the labor power of humans. And these demands on part of the technosphere, as he sees it, are just the beginning of an accelerating process: as it grows in size, it feeds back into an ever-increasing demand for more energy. Put in teleological terms, the technosphere wields its own will, operating only in accordance with its own reproduction. As Haff imagines it, the technosphere thus constitutes a challenge to the anthropocentric tendency to put humans at the center of things. Of course, on some level, the environmental change we witness in its wake is caused by humans, but there are many

31 Ibid., 127.
32 Ibid., 302.

systems in the world that embed humans in such a way that it is difficult to distinguish what is strictly human about them and what "human" in this context even means. In short, the Cartesian concentration of agency in the human subject — constitutive of the illusion of control — simply does not hold anymore.

In an ironic twist of fate, then, the key conceptual contribution of the technosphere is perhaps also its chief weakness. Although the term may be useful in addressing the idea that ecosystems are dynamic and do not have a natural stability, but that they are persistently altered and worked upon by organisms, a number of earth scientists have argued that this means that we ought to give up on the nature/artifice dichotomy that the concept of the technosphere operates on. Even though the geographer Erle Ellis agrees that understanding technological processes is as crucial as understanding our planet's biological and geophysical processes, he nonetheless maintains that any description of our situation that stresses the artificiality of the earth starts out from the wrong assumption. Since novel ecosystems point precisely to the boundary problem between natural and artificial systems, we ought not to draw the conclusion that the earth is therefore becoming increasingly artificial, but rather that technological modification is natural.[33] "In this sense," Haff admits, "one might say that technology is the next biology."[34] Whether or not we find value in the technosphere as a discrete concept, then, is partly contingent on what we define as "technology," and whether we believe that all organisms — including humans — have always been in the business of altering their surroundings, and whether natural/artificial hybrids are anything new.

The technosphere merely serves to further illustrate the tension within the Anthropocene narrative: it is not as straightforward as it might seem at first glance, because if there is anything that earth system scientists agree on, it is that the environmental conditions for human flourishing are not strictly "natural"

33 Ellis, "Ecology in an Anthropogenic Biosphere."
34 Haff, "Technology as a Geological Phenomenon," 302.

and in fact never have been. Since prehistory, humans—just like every other organism—have engineered their environment to suit their own survival. In fact, this insight is crucial to the very definition of an "ecosystem": the interaction between opposing forces in such a manner that the system tends toward certain temporarily stable states. On the one hand, there is no clearer illustration of the extent of humanity's effect on the earth than the fact that the preservation of species and the stability of ecosystems is dependent on an ever-increasing human involvement.[35] On the other hand, this is also its irony: nature is possible only insofar as a growing scale of human enterprise maintains its pristineness through artificial conservation practices. Human activity within the Anthropocene thus names the kind of organization that has come to unsettle the earth system out of which it has emerged, in the double-edged sense that it both poses a threat to the previous order and also embodies the capacity for progress and creative evolution. If we are to honor the essential features of nature, which means acting in accordance with our own human nature, then we ought to reconceptualize risk into opportunity. Continued production and artificial alteration, it turns out, is in fact our *natural* mode of existence.

§

The Anthropocene is therefore not a strictly scientific problem, but is also *philosophically* interesting in several ways, one of which is that it is deeply implicated with our cultural understanding of the distinction between the natural and the artificial. When humankind itself becomes a natural force—or, when that which we previously perceived as natural is revealed to be increasingly human-made—then ontological dichotomies such as nature/artifice and subject/object no longer seem to function in their accustomed fashion. Along with the aforementioned practicality of these categorizations, the conventional understanding of disciplinary methods for the produc-

35 Vitousek, "Human Domination of Earth's Ecosystems," 499.

tion of knowledge — the natural sciences, on the one hand, and the human sciences, on the other — seems to have reached its limit too. The conjoined mixture of a rapidly changing climate, an expanding industrial metabolism, a growing impact of global land-use change, and a scaling up of urbanized environments demonstrates that questions about the relationship between cause and effect, means and ends, and quantity and quality require us to face a whole range of philosophically rich topics. A sense of amazement at the wonder of the earth has thus arisen once more, one that seeks to make sense of its own historical conditions and forms of expressing itself: How can we know? What can we do? And to what extent are these two questions (in)separable from each other? With what means, methods, and senses can we encounter a planet with which our own activities are intricately entangled?[36] One of the most central questions of human civilization — our place and role in the earth's evolution — consequently remains a fundamental concern. Is *Homo sapiens* merely *primus inter pares* — no more than first among equals in the animal world — or does our technical ability to radically alter nature set us apart from other species in an essential way? Are we guardians or tenants of the earth? Are we, thanks to the global reach and range of technological power, finally masters of our own destiny, or does the global proliferation of technical systems indicate an increasingly path-dependent future? These are, certainly, fundamental questions of philosophy, but the unprecedented capacity of our computational and information technology, and our remarkable ability to manipulate our surroundings on a global scale, have reached a point where we are led to ask such perennial questions anew,[37] which means that we are likely to find both historical bearing and contemporary resonance when doing so.

36 Bernd M. Scherer, "A Report: An Introduction," in *The Anthropocene Project: A Report*, ed. Bernd M. Scherer (Berlin: Haus der Kulturen der Welt, 2014), 4.

37 See, for instance, Benjamin H. Bratton, *The Terraforming* (Moscow: Strelka Press, 2019).

As technology keeps transforming the conditions by which we understand and interact with the world around us, it becomes relevant to ask questions about the ontological aspects of these patterns of change. Although it might be something of a cliché, it is no less true to say that modern technological developments have transformed the very foundation of our understanding of what it means to be human. The ontological boundary between human and machine is constantly called into question as information technology, genetic engineering, and artificial intelligence continue to develop at a bewildering pace. What we previously understood as the defining properties of the human—agency, consciousness, affect, and the very operation of reason—now seem to be inextricably bound up with technically replicable processes. Technology, to put it crudely, appears more and more as an ontological state rather than an instrument of a priori reason, representing something that no longer resembles merely a collection of prostheses, and that would thus be merely supernumerary or supplemental to human nature, but more akin to the most basic and enabling feature of the human condition. This is no less true for concerns about environmental change. It is in the interaction between humans and their environment, and how this understanding is enacted, challenged, and renegotiated, both within the sciences and in our broader culture, that we come up against the pressing issue of anthropogenic environmental change. Climate change, biodiversity loss, pollution, and exploitation of natural resources have contributed to making this relationship more relevant than ever. These are examples of problems that neither the natural nor the human sciences have a monopoly on dealing with, nor the capacity to solve on their own. Facing up to our contemporary environmental problems requires an understanding of the behavior of physical processes in the natural world, but also of the conceptions of "nature" and "artifice" that circulate in such narratives.

If the humanities are relatively new to the party, earth scientists have already worried for the last four decades that humankind, in its effort to extract greater and greater use value, has come to disrupt the biospheric conditions upon which it depends, and in so doing has inadvertently caused substantial shifts to the very structure of the earth system. The concern that these scientists now have about global environmental change arises from our growing awareness that ecosystem services are interdependent and that, on a planetary scale, their long-term supply is at risk if supporting and regulating processes are put under too much stress. Through the power of global technology, processes with unexpected, large-scale, and domino-like ramifications may be set in motion by qualitatively shifting the conditions of what was previously thought of as a natural order. It increasingly seems like an impossible task to predict the totality of the effects of our actions into an indeterminate future precisely because what we used to regard as "constants" have already undergone qualitative changes. Moreover, such an insight makes it equally awkward to posit a constant or unchangeable human nature. Alongside unlocking radical opportunities to alter both ourselves and our environment, this poses a novel risk for our species. The golden promises of technological progress have turned into a threat by inadvertently undermining the very foundation for our existence, meaning that we no longer have control over nature through technology—the prerequisite for the scientific and industrial revolution—insofar as we cannot know in advance the full consequences of our actions or their damaging effects on the natural world or on future generations. Rather, with its capacity to qualitatively alter nature, technology seems to be at a risk of making itself autonomous and taking control over us instead. Today, the human condition itself has become the subject of technical reshaping.

However, to concede that global technology obscures our responsibility is not the same as to say that we are therefore exempt from moral obligation. On the contrary, because of the power humans now wield, these ethical questions remain more crucial than ever. Even if the destruction of the entire biosphere remains unlikely, it could still shift to new modes of interaction within which there may eventually be no room for humankind—causing immense suffering and death for humans and nonhumans alike in the process. Nevertheless, the point is that mere calculation is not enough. To claim that all we need is better and more powerful computers to increase the precision of our models is to belie the full complexity of the problem. Regardless of our calculative capacity, what we face is equally a problem of qualitative character. In "Philosophy at the End of the Century" (1994), the philosopher Hans Jonas described the crisis he understood to be fundamental to the existential risk of global environmental change as one that requires us to return to "one of the oldest philosophical questions, that of the relationship between human being and nature, between mind and matter — in other words, the age-old question of dualism."[38] In the modern scientific effort to describe and explain nature objectively, the scientist is compelled to subtract the secondary qualities of the subject — his purposes, emotions, and interests — in order to demonstrate that from the point of view of the scientific method it is enough to refer to causal connections. As Jonas stresses, however, the effort to explain nature is of an entirely different character than the effort to understand it. There is, in other words, an ontological locus in nature, which prevents scientific methodology from determining what nature *is*. Nature cannot be reduced to knowledge about nature, or, in other words, to nature as known by the natural sciences.

In the same way, conceding to the prevailing disagreements within the geoscientific community on even the most basic methodological questions regarding the Anthropocene — which stratigraphic, atmospheric, and biotic variables, for instance, should take precedence in establishing its onset, how significant a change in value of these variables should be expected, and whether the transition should be tracked on a global or regional scale of analysis[39] — proponents of the concept have argued that "to assign a more specific date to the onset

38 Hans Jonas, "Philosophy at the End of the Century: A Survey of Its Past and Future," *Social Research* 61, no. 4 (1994): 826.
39 Bruce D. Smith and Melinda A. Zeder, "The Onset of the Anthropocene," *Anthropocene* 4 (2013): 8.

of the 'anthropocene' seems somewhat arbitrary."[40] The point is that the conceptual power of the Anthropocene does not fall exclusively within the ontic realm of factual thinghood, but also in the way it challenges the ontological distinction between natural and human history. Regardless of whether the industrial revolution, or earlier alterations within the Holocene, have left unambiguous geological signals of human activity that are synchronous around the globe,[41] this question is not enough on its own to discredit the concept's analytical value. Instead, its contribution lies precisely in questioning the entire dichotomy that proceeds from an essential separation between the artificial products of humankind and the natural products of the earth. Consequently, in the attempt to identify a date to mark the beginning of the proposed geological time unit, disagreements within the geoscientific community have concerned not only how anthropogenic effect should be measured, but even more fundamentally what "anthropogenic" even entails. As Simon Lewis and Mark Maslin have stressed, what is at stake in the formal definition of the Anthropocene is not limited to the status of an empirical fact — one that happens to be of specific interest to earth scientists — but equally concerns ontological questions about the essence of technology and of human nature:

> Defining an early start date may, in political terms, "normalize" global environmental change. Meanwhile, agreeing a later start date related to the Industrial Revolution may, for example, be used to assign historical responsibility for carbon dioxide emissions to particular countries or regions during the industrial era. More broadly, the formal definition of the Anthropocene makes scientists arbiters, to an extent, of

40 Crutzen and Stoermer, "The 'Anthropocene,'" 17. See also Todd J. Braje and Jon M. Erlandson, "Looking Forward, Looking Back: Humans, Anthropogenic Change, and the Anthropocene," *Anthropocene* 4 (2013): 116–21.

41 For a survey of methodological disputes pertaining to the Anthropocene, see Richard Monastersky, "Anthropocene: The Human Age," *Nature* 519, no. 7542 (2015): 144–47.

the human–environment relationship, itself an act with consequences beyond geology.[42]

In the philosopher Clive Hamilton's words, "the appearance of this new object, the Earth System, has ontological meaning. It invites us to think about the Earth in a new way, an Earth in which it is possible for humankind to participate directly in its evolution by influencing the constantly changing processes that constitute it. It therefore brings out the conception of a joint human-earth story."[43] For Hamilton, the Anthropocene indicates that humankind is no longer ontologically distinguishable by some fundamental essence that constitutes its exceptionality. Rather, the human has become incorporated into the immanence of an unqualified immersion or embeddedness in the complex processes of geophysical flows and folds, bringing humans "down to earth," so to speak. The concept of the Anthropocene is thus not isolated to the scientific concerns of geology, climate science, or even earth system science, but moves beyond disputes over empirical evidence insofar as it more generally "represents a ground-breaking attempt to think together Earth processes, life, [and] human enterprise [...] into a totalizing framework."[44] Such a proposed convergence of human enterprise with earth processes is philosophically relevant since it renders nature and artifice symmetric, in the simple sense that artificial processes of change are made to appear in the same ontological register as those of natural processes of change. By implication, artifice is not merely considered to play a supplementary role in relation to nature, nor is reason depicted to appear on earth as a manifestation of something superlunary or

42 Simon L. Lewis and Mark A. Maslin, "Defining the Anthropocene," *Nature* 519, no. 7542 (2015): 171.

43 Clive Hamilton, *Defiant Earth: The Fate of Humans in the Anthropocene* (Cambridge: Polity Press, 2017), 21.

44 Clive Hamilton, Christophe Bonneuil, and François Gemenne, "Thinking the Anthropocene," in *The Anthropocene and the Global Environmental Crisis: Rethinking Modernity in a New Epoch,* eds. Clive Hamilton, Christophe Bonneuil, and François Gemenne (London: Routledge, 2015), 2.

transcendental, but rather *as* earth, as intrinsic or immanent to its self-organizing efforts.[45]

Now, the message of the earth system paradigm is clear: nature can no longer be understood as a domain separate from human artifice. In the Anthropocene, we have seemingly concrete technological production giving rise to shifts in the qualitative properties of the earth system, which alter the conditions for human existence and give rise to new patterns of organization. The modern dichotomy between nature and artifice has imploded, we are told, resulting in a deep intertwining of the fates of humanity and the earth. Zalasiewicz states that "the Anthropocene represents a new phase in the history of both humankind and of the Earth, when natural forces and human forces became intertwined, so that the fate of one determines the fate of the other."[46] As most forcefully argued by the historian Dipesh Chakrabarty, the Anthropocene entails a constant conceptual circulation across deep and historical time, posing a powerful challenge to the modern ontological separation between the two:

> The distinction between human and natural histories — much of which had been preserved even in environmental histories that saw the two entities in interaction — has begun to collapse. For it is no longer a question simply of man having an interactive relation with nature. This humans have always had, or at least that is how man has been imagined in a large part of what is generally called the Western tradition. Now it is being claimed that humans are a force of nature in the geological sense.[47]

---

45 Jochem Zwier and Vincent Blok, "Seeing through the Fumes: Technology and Asymmetry in the Anthropocene," *Human Studies* 42 (2019): 623. See also Jochem Zwier and Vincent Blok, "Saving Earth: Encountering Heidegger's Philosophy of Technology in the Anthropocene," *Techné: Research in Philosophy and Technology* 21, nos. 2–3 (2017): 222–42.

46 Zalasiewicz, "The New World of the Anthropocene," 2231.

47 Dipesh Chakrabarty, "The Climate of History: Four Theses," *Critical Inquiry* 35, no. 2 (2009): 207.

By "the geological sense," Chakrabarty refers not only to the empirically observable effects that humans have had on the earth, but also to the alterations in self-consciousness undergone by a culture experiencing an increasing unease as it pertains to its own alienation, that is, as an active participant in and as a driving force of global environmental change. Rockström and Steffen have repeatedly emphasized that "we are the first generation with widespread knowledge of how our activities influence the Earth system, and the first generation with the power and the responsibility to change our relationship to the planet."[48] It seems that, if anything, the Anthropocene signals that there is now an encouragingly widespread recognition that we are in the midst of a unique phase in human history, where, for the first time, we have been made aware of the causal connection between events on the geological scale and the everyday practices of our daily lives. It challenges the assumption that whatever remains natural about humankind — its essence — has no real history, while the rest of the world supposedly belongs to the province of an entirely distinct "natural history," such that, insofar as humankind does have a history, it is only relevant to the extent to which its activities are artificial and thus unnatural. History, in the conventional sense of the word, and to which artifice thus belongs, only commenced, then, when humans began to act "unnaturally" — to cultivate crops, craft tools, and eventually erect entire civilizations.[49] According to Chakrabarty, our knowledge of anthropogenic environmental change has breached this once seemingly impregnable ontological wall of separation, and the Anthropocene, for him, is precisely the name for the existential implications of this ontological collapse.

---

48 Will Steffen et al., "The Anthropocene: From Global Change to Planetary Stewardship," *Ambio* 40, no. 7 (2011): 757.

49 Timothy J. LeCain, "Heralding a New Humanism: The Radical Implications of Chakrabarty's 'Four Theses,'" in *Whose Anthropocene? Revisiting Dipesh Chakrabarty's "Four Theses"* — RCC *Perspectives: Transformations in Environment and Society,* eds. Robert Emmett and Thomas Lekan (Munich: Rachel Carson Center for Environment and Society, 2016), 2:15.

What we have supposedly come to realize through the Anthropocene, then, is that our existence as humans does not reside in some isolated and unchangeable substance but rather in our interaction with other natural forces, which means that what we perceive as a subject of political and ethical relevance may look very different depending on the spatial and temporal scales that we select for observation. The innovativeness of the Anthropocene as a concept, according to its proponents, lies precisely in its insistence on situating humans and their actions within the large-scale structure of the earth as a whole, that is, within planetary assemblages that emphasize the fluidity, exchangeability, and multiple functionalities of systems and their connectivity, and in whose midst "humankind" and "technology" constitute but certain constellations in the fractal geography of the earth system.[50] According to Hamilton, the implications of the Anthropocene cannot be reduced to the broadening impact of humans on the natural world, which merely extends what has been an ongoing process for centuries or millennia. Instead, what this new geological epoch supposedly denotes is a shift in focus, much like the shift from the interest in the early science of thermodynamics on phenomena taking place at or near equilibrium in energetically isolated systems to the developments of the second half of the twentieth century with the rise of far-from-equilibrium thermodynamics and an accompanying interest in the evolution of dissipative structures in systems open to energy fluxes. In short, the question of the human has become a topological question: a problem of localization.[51] We should no longer try to determine *who* or *what* the human essentially is, but *where* it is in relation to its environment. With the transition from anthropological substance to

50 Clive Hamilton, "Define the Anthropocene in Terms of the Whole Earth," *Nature* 536, no. 7616 (2016): 251.

51 Peter Sloterdijk, *Bubbles — Spheres*, vol. 1: *Microspherology*, trans. Wieland Hoban (Los Angeles: Semiotext(e), 2011), 630. See also Pieter Lemmens and Yuk Hui, "Reframing the Technosphere: Peter Sloterdijk and Bernard Stiegler's Anthropotechnological Diagnoses of the Anthropocene," *Krisis* 2 (2017): 28–31.

anthropological function—from *natura hominis* to *conditio humana*—humankind's way of being emerges as a constellation, leading to the confounding irony that as humanity has become powerful enough to enter the scene of geological time, its concrete subjectivity has concomitantly been questioned.[52]

Consequently, in the face of global environmental change, claims about the essence of technology have not only been diverse but sometimes even contradictory: precisely when the power of technology has become recognized as an existential danger because of its capacity to alter the environment globally, technological production and alteration has come to be regarded as natural processes.[53] Put slightly more provocatively, within the Anthropocene discourse there is virtually no question of nature since it is merely a product of technology; nor is technology any longer merely prosthetic, meaning mere artificial replacement or supplement. Instead, the order of things seems to have become inverted: technology is not only supplementary, but rather it has become the ground in contradistinction to the figure. However, this raises a crucial question: How are we to make sense of the radically incommensurable

---

52 For an overview of this confounding irony, see Arianne Conty, "Who Is to Interpret the Anthropocene? Nature and Culture in the Academy," *La Deleuziana* 4 (2016): 19–44.

53 My concern here is basically an inverted variant of the same argument made by the literary theorist Timothy Morton. Whereas I am interested, herein, in the question concerning technology, Morton, on his end, has for a long time been engaged in a sustained critique of the concept of nature. He asks himself, "At what point do we stop, if at all, drawing the line between *environment* and *non-environment:* The atmosphere? Earth's gravitational field? Earth's magnetic field, without which everything would be scorched by solar winds? The sun, without which we would not be alive at all? The Galaxy? Does the environment include or exclude us? Is it natural or artificial, or both? Can we put it in a conceptual box? Might the word *environment* be the wrong word? *Environment,* the upgrade of *Nature,* is fraught with difficulty. This is ironic, since what we often call the environment is being changed, degraded, and eroded (and destroyed!) by global forces of industry and capitalism. Just when we need to know what it is, it is disappearing"; *The Ecological Thought* (Cambridge: Harvard University Press, 2010), 10.

insights of the Anthropocene that global technology constitutes a profoundly destabilizing force on nature, and that technical processes of transformation have always been, and always will continue to be, perfectly natural? How are we to begin to make sense of the claim that technology is, in some sense, not an aberrant condition but really our natural state? And at the same time as it is recognized that it may ultimately destroy us?

§

Calling attention to the twofold meaning of technology, the philosopher Jean-Luc Nancy has argued that the concept of the technological, as it figures in the Western tradition, has since its very inception been characterized by an inherent ambiguity: technology simultaneously *supplements* and *supplants* nature.[54] Indeed, to speak, as the Western tradition continuously did, of a twofold meaning of technology, is to invoke a conceptual layer of intellectual history that runs from Plato and Aristotle to the present. Although technology, in Western philosophy, has always been inscribed in the natural world in one way or another, the uneasy and often inconsistent relationship between artifice and nature has had an enduring influence upon our conception of humankind as exceptional in the animal kingdom, residing, and paradoxically so, simultaneously inside and outside of nature. The Greek myth of Prometheus's theft of fire from the gods and the Judeo-Christian myth of the fall of humankind — both of which proceed from a prelapsarian state wherein all living beings possessed a prescribed role in the normative order of nature as decided by the divine will — are some of the most foundational mythologies to have shaped the cultural self-consciousness of the Western tradition. In each, the origin of humanity represents two distinct moments. On the one hand, a moment of creation as merely an animal within the

54 Jean-Luc Nancy and Aurelien Barrau, *What's These Worlds Coming To?*, trans. Travis Holloway and Flor Méchain (New York: Fordham University Press, 2015), 42.

natural order, albeit with a duality that distinguishes humans from other animals, such as a negative absence of innate animal capacities, or a positive resemblance to the divine. On the other hand, a moment of rupture, as humankind is wrenched from the natural order only to be related to it anew. Whereas in the myth of Prometheus it is the acquisition of practical knowledge that wrenches humans from nature insofar as it empowers them to undermine this order by carving out their own place within it, in the myth of the biblical fall it is similarly the acquisition of theoretical knowledge that enables such a transgression since it is only on the basis of understanding that it is possible for humans to subvert it. To be sure, there are numerous other aspects of these mythologies that have influenced the development of humanism, from the human ideal as unmarked by animal traits to the persistence of the sexual specificity of original sin. But it is nonetheless the fundamental generality of the relationship between creation and rupture that we find center stage: the human is but one being among many in the order of the natural world, yet it is simultaneously placed in a unique relationship to nature as a whole, with, at least in principle, the capacity to understand and exploit it. It is in the light of this generality that humans are said to resemble the divine, or to have stolen something from it, such that the subsequent theological negotiation between humankind and God becomes a matter of curtailing the scope of human actions or of folding humanity's rupture back into the divine order again, so that the range of the practical and theoretical capacities of humans is safely circumscribed by their natural role.[55]

When it comes to the history of Western metaphysics, it is most often Aristotle who is recognized as having instigated an ontology of the artifact. From Aristotle onward, technol-

55 Peter Wolfendale, "The Reformatting of *Homo Sapiens*," *Angelaki: Journal of the Theoretical Humanities* 24, no. 1 (2019): 55–66. For a more detailed interpretation of the Promethean myth from the point of view of a philosophy of technology, see Bernard Stiegler, *Technics and Time*, vol. 1: *The Fault of Epimetheus*, trans. George Colins and Richard Beardsworth (Stanford: Stanford University Press, 1998), 185–203.

ogy—or *technē*, the ancient Greek word for "craftsmanship" or "art"—has conventionally been understood as an essentially inert and neutral instrument whose status is entirely determined by the use to which it is put by humans. Whereas artificial products are generated solely by external causes—namely, by an intention external to the object itself—natural products, such as animals and plants, are animated by an inner, final cause, that is, they move, grow, change, and even reproduce themselves on their own, driven by a purpose of nature and not of humankind. Nature contains the principle of its own motion—an organism will grow up, mature, and wither away all by itself—but the fabricated artifact requires an efficient cause, such as an artificer, to bring it into being, alter it, or recycle its materials into something new. Artifacts, in other words, cannot organize themselves, because without external care and intervention they neither come into being nor persist but slowly deteriorate and vanish by losing their artificial forms and decomposing into raw materials. Such is the understanding of the essence of technology that has had a huge influence on Western metaphysics ever since: technology is a prosthesis that must be considered as a supplement to nature, reason, or the human; an instrument that can be utilized for good or ill depending on the intention of the artificer who crafts or wields it. Insofar as the production process of *technē* takes place in a manner directly analogous to natural process of generation, artificial products are but imitations of natural products.[56]

However, there is already another metaphor hiding in the Greek word for "instrument," *organon,* which is only made all the more apparent once one recognizes that the organism, in Aristotle's work, is resembled by an artificer and its tools. Indeed, it is not an accident that the vital parts of the organism are called precisely "organs," for Aristotle saw a close relationship between the two areas. In fact, Aristotle frequently sets the organ and craft models side by side, such that the craft model is made to

56 Aristotle, *Aristotle's "Physics,"* ed. William D. Ross (Oxford: Clarendon, 1936), 351 (bk. 2, 194a21).

shed light on the function of the living body, all the while craft itself is described in terms of an organic process. In this sense, we are allowed to understand one of the areas with the help of the other, and in such a manner that they together bring clarity to what distinguishes the more abstract question of the production of things in general. In his *Metaphysics*, Aristotle writes about the natural and the artificial fabrication of things more specifically. An organism, he argues, develops naturally because both form and matter are present from the beginning. Organic matter is thus inconceivable without form, for the organic process is always guided by it, namely, the shape of the fully developed organism. When, on the other hand, an artificer produces an artifact, it has as the starting point for its business raw materials, that is, matter without an accompanying form. The form he strives for in his work does therefore not reside in the matter but exists solely in his own mind.[57] This kind of artificial production means that he is able to grant matter a form thanks to his professional skills. The professional skill, *technē*, thus unites those who can produce something out of matter that is not already there in advance. In fact, when Aristotle coins the concept of *húlē* in order to describe what we today understand by the Latin concept of *matter*, he proceeds precisely from the craft model. But the same concept also has an immediate proximity to the organic world, and associates with the same self-assuredness to the domain of biological growth as it does to the domain of craftsmanship, which means that Aristotle can move between these two domains rather effortlessly. For Aristotle, as for his contemporaries, it was evident that one could understand the living and the crafted in a similar way. Technology in the modern era came to be associated with mechanical processes, but the ancient Greeks saw the interaction between the artificer and its tools in analogy with the coordination of the organs, the "tools" or "instruments," of the living body. Although the implements in question were taken as mechanical in separation from each

57 Aristotle, *The Metaphysics: Books I–IX*, trans. Hugh Tredennick (London: William Heinemann Ltd., 1933), 339, 341 (bk. 7, 1032b).

other, in the hands of the artificer they were nonetheless perceived as subordinated to its conscious intention, and just like in the body they are thus put to work in relation to the function provided by its form.[58] Despite the secondary status granted to *technē*, then, Aristotle simultaneously observed that it perfects that which nature by itself is unable to achieve,[59] which suggests that technology may play a much more fundamental role than the superficial status of a mere imitation would have us believe. This is the way in which technology, by proceeding analogously to nature, on the one hand, brings it to perfection, on the other.

But even if the Promethean myth and the Aristotelean ontology of the artifact points to a prevailing cultural self-consciousness in Western history that refers to the twofold meaning of technology, there is nevertheless the sense that we are faced with an entirely new situation in the wake of modernity. Premodern modes of production, limited by materials and energy sources given in nature, were unable to introduce products and processes at odds with the biospheric conditions of humanity's terrestrial existence. Yet, with the introduction first of steam, then electrical, and later nuclear power, along with the speed of ever-expanding industries for the mass production of novel compounds—plastics, synthetic pesticides, immunomodulatory agents, and so on—and infrastructures for the mass distribution of goods in space and time, such is no longer the case. Already present in ancient Greek philosophy, the uneasy tension inherent in the term "technology"—as that which at once passively brings nature forth and actively intervenes into it—can arguably be said to properly have come to the fore first in the modern period. Not the least because we are presented, in modernity, with an increasing number of such tensions, on a microscopic and a macroscopic scale. At one end of the spectrum, with the advent of chemical, biotechnical, and genetic engineering, it is no longer clear where the natural ends and the

58 Ibid., 337, 339 (bk. 7, 1032a).
59 Aristotle, *Aristotle's "Physics,"* 357 (bk. 2, 199a15).

human-made begins.[60] Chemically engineered polymers unify form and matter at the level of atomic structure and can now exhibit stability equivalent to — or even greater than — that of natural compounds. Simultaneously, at the macroscopic end of the spectrum, we seem to be faced with what can be best described as the globalization of technology. Humans now move more rock and soil than all of the earth's glaciers and rivers combined, fix more nitrogen than microbial activity does, and consume vast quantities of resources. Influenced by human activity not just in part but as a whole and on the verge of being taken over by global technological systems, has not even the earth itself become artificial?

In the late 1980s, the environmentalist Bill McKibben argued that in the wake of anthropogenic global change, the concept of nature as a "separate and wild province, the world apart from man to which he adapted, [and] under whose rules he was born and died"[61] had itself perished. Such an increase in magnitude of human alteration raises questions about the point at which natural products become technological products, or, more generally, the point at which nature turns into artifice. The issue here might be one of degree, but it also has qualitative consequences. McKibben writes:

> This new rupture with nature is different not only in scope but also in kind from salmon tins in an English stream. We have changed the atmosphere, and thus we are changing the weather. By changing the weather, we make every spot on earth man-made and artificial. We have deprived nature of its independence, and that is fatal to its meaning. Nature's independence is its meaning; without it there is nothing but us.[62]

60 Carl Mitcham, *Thinking through Technology: The Path between Engineering and Philosophy* (Chicago: University of Chicago Press, 1994), 172–74.
61 Bill McKibben, *The End of Nature* (New York: Random House, 1989), 48.
62 Ibid., 58.

Ironically, technical artifacts, precisely insofar as they begin to overtake nature, also begin to appear as increasingly indistinguishable from it. As the anthropologist Bruno Latour has observed,[63] the modern effort to purify nature from artifice has not led to the resolution of the ancient Greek tension in the ontological sanitation of substance dualism but, quite to the contrary, to its dissolution in an indifferent monism that stresses their fundamental contamination as "technonature."[64] Latour states, "The Earth is no longer 'objective'; it cannot be put at a distance and emptied of all its humans. Human action is visible everywhere — in the construction of knowledge *as well as* in the production of the phenomena those sciences are called to register."[65] Instead of conforming to the dualist ontology of Western modernity, Latour finds that, on the contrary, the modern era points in the complete opposite direction, toward an increased unification of human activities with the rest of nature through what the biologists Humberto Maturana and Francisco Varela called "the unbroken coincidence of our being, our doing, and our knowing,"[66] that is, the active implication of humankind in the production of nature as such. In other words, there is a sense in which the earth, as an object of study, cannot simply be taken as an object. It can no longer be viewed as if from a distance, separate from the activities of all its human inhabitants. Our efforts to predict nature in order to domesticate and control it also tend to change it in unpredictable ways, because to measure, to represent, and to compose the shape of the earth is essentially to study an entity to which we ourselves are intimately bound. What distinguishes our current condition from

---

63 Bruno Latour, *We Have Never Been Modern*, trans. Catherine Porter (Cambridge: Harvard University Press, 1993).

64 Damian F. White and Chris Wilbert, eds., *Technonatures: Environments, Technologies, Spaces, and Places in the Twenty-First Century* (Waterloo: Wilfrid Laurier University Press, 2010).

65 Bruno Latour, "Agency at the Time of the Anthropocene," *New Literary History* 45, no. 1 (2014): 5–6.

66 Humberto R. Maturana and Francisco J. Varela, *The Tree of Knowledge: The Biological Roots of Human Understanding*, trans. Robert Paolucci (Boston: Shambhala, 1987), 25.

the ancient Greeks is that once the Aristotelean tension inherent to technology comes to the fore in modernity, it is immediately dissolved in antinomy.

But as issues of environmental change suggest, such a ubiquity of technology in nature may not be particularly helpful when it is precisely the question of the technical alteration of nature by humans that is at stake. With the essential indeterminacy between nature and artifice, the question of their relationship is also abolished, thereby leaving us with a critical dilemma: If nature is intertwined with artifice in its very being, then how can we even begin to imagine a different attitude to the environment than our current, technological one? For in the undecidability between nature and artifice, with the artificialization of nature we are simultaneously faced with the naturalization of artifice. So, what happens to the Western canon once technology becomes so ubiquitous that it can be said to have become a global phenomenon, to the point that it not only permeates nature throughout but dissolves the very tension that has characterized the relationship between nature and technology ever since Aristotle? This book, far from attempting to exhaustively answer this question, concentrates on a shift in the history of earth science as it pertains to the question concerning essence of technology.

The Question Concerning Planetary Technicity

In September 1966, as the German philosopher Martin Heidegger entered the last decade of his life, he gave an interview to be published posthumously by the magazine *Der Spiegel,* wherein he warned precisely about the eclipse of the natural by the artificial. According to Heidegger, the phenomenon we call "globalization" demonstrates that a purely calculable understanding of being has been extended to perfectly encompass the earth, such that our relationship to our terrestrial abode has exclusively come to take on what he called the "form of planetary technicity."[67] This particular mode, which is a mode of disclosure, names the colonization of the earth by scientific modernity, and marks the destruction of all limits to the instrumental project of technological manipulation. Consequently, technology is disclosed as everywhere and nowhere: there are hardly any pristine environments anymore, to the degree that "pristine" has lost its meaning; anthropogenic environmental change has become so prevalent that it is conceived of as a literal "force of nature"; and global monitoring of the entire earth has established a panoptic view from nowhere, where only the monitoring instruments themselves remain out of sight. For Heidegger, "planetary technicity" is thus the name for an ontological structure that sets up a general equivalence of beings wherein the very distinction between artifice and nature loses its pertinence, in such a way that the former is disclosed not as an imitation of the latter but rather as its originary revelation. In this sense, there is no natural environment that would not be open to technical supplements, and even more radically, no natural environment that is not already artificial.[68] Proceeding from the notion of a radically interconnected planet and consequently an in-itself artificial earth, narratives of technological globalization establish a discursive framework within which the increased technical manipulation of nature appears as an inevitability, and in such a manner that responsibility is ultimately reduced to a task for technocrats to compile intricate quantitative estimates for the most efficient use of the planet's resources.

Indeed, if the label "Anthropocene"—a combination of the Greek word for "humankind," *anthrōpos,* with the word *kainos,* meaning "recent" or "new," referring to a geological epoch whose novelty is defined by the global environmental impact of humanity—seems to undermine its own prefix, this would have been entirely unsurprising to Heidegger, who, already as

---

67 Martin Heidegger, "'Only a God Can Save Us': The *Spiegel* Interview," in *Heidegger: The Man and the Thinker,* ed. Thomas Sheehan, trans. William J. Richardson (London: Transaction Publishers, 1981), 55. In the original version of the interview, Heidegger uses the term *Planetarische Technik.*

68 Nancy and Barrau, *What's These Worlds Coming To?,* 46–47.

early as 1939, not only proposed that the modern era of industrial conquest was leading toward the enframing of the entire earth in the form of the sum total of a standing reserve, to be ordered and organized by the hegemony of instrumental reason, but also anticipated that the global reign of instrumentalism would inevitably expose humans to the same objectifying gaze and reduce them to mere objects to be technically manipulated. In fact, Heidegger located this antinomy at the heart of planetary technicity: it describes the condition in which historical humankind experiences the sensation of unlimited power, yet at the same time a lack of meaning and a sense of existential desolation. Doubts and anxieties about ourselves live side by side with a crude fanaticism and blind faith in technological progress. Hope mixes with fear, obscurantism with rationalism, and sentiments of fundamental powerlessness with a planetary will to power. In the wake of modernity, technology is seen to overrun and command the entire globe. It figures as an image of immensely distributed and complex infrastructures — networks of technical apparatuses, modes of organization, and so on — and its entry onto the global scene suggests a state of total dominion: the setting up of planetary-wide technological systems that objectify the earth and all of its entities in accordance with a global technocratic framework of imperial control. It is in this vein that Hans Joachim Schellnhuber, the founding director of the Potsdam Institute for Climate Impact Research and one of the most outspoken champions of earth system science, advocates "the emergence of a modern 'Leviathan,' embodying teledemocracy and putting the seventeenth-century imagination of the English philosopher Thomas Hobbes into the shade."[69] In fact, it is significant how closely earth system concepts such as planetary boundaries and thresholds resemble the Hobbesian justification of the Leviathan — or, to be more precise, its rebirth in environmentalist disguise. For these restrictions are often justified by reference to resources. It has become a topic so central

69 Hans J. Schellnhuber, "'Earth System' Analysis and the Second Copernican Revolution," *Nature* 402, no. 6761 (1999): 22.

to discussions about global environmental degradation that it seems almost unimaginable to conceive of an environmentalist discourse uncoupled from quantitative worries about diminution or excess, such that even our best efforts to maximize the benefits and minimize the costs of trading economic growth for ecological well-being — the restoration of natural capital, fostering ecosystem services, and such — are first and foremost concerned with an ever-widening scope of instrumental calculation and management, wherein humanity too is ultimately subsumed in the biopolitical parlance of ecosystem dynamics, ecoregional constituencies, and population stability and sustainability.[70]

On the other hand, concerns for the global reach of technology also gesture toward a sense of scale that far exceeds human control. In their efforts to render the entire globe governable, humans experience themselves being helplessly swept away by the very technological systems that they initially conceived to be in their possession. According to Heidegger's diagnosis of the modern condition, this existential anxiety has its roots in an unhealthy obsession with the general planning of beings, such that everything is viewed in its functional aspect, which causes us to lose sight of what is sacrificed in the mobilization of beings for goals that remain ultimately obscure. Because humans are objectified and instrumentalized into nothing but a means for planetary-scale management, Heidegger finds that the only kind of humanity that is capable of the unconditional completion of such a nihilism is a humanity for whom the question of the meaning of being has become forgotten, and in whose wake there is no other option but to perceive ourselves in the midst

---

70 On the biopolitics of the earth system paradigm, see Ola Uhrqvist and Eva Lövbrand, "Rendering Global Change Problematic: The Constitutive Effects of Earth System Research in the IGBP and the IHDP," *Environmental Politics* 23, no. 2 (2014): 339–56. On the biopolitics of the discourse of global environmental change more broadly, see Paul Rutherford, "The Entry of Life into History," in *Discourses of the Environment,* ed. Éric Darier (Oxford: Blackwell, 1999), 37–62, and Timothy W. Luke, "On Environmentality: Geo-Power and Eco-Knowledge in the Discourses of Contemporary Environmentalism," *Cultural Critique* 31 (1995): 57–81.

of those inhuman forces — earthquakes, flash floods, tsunamis, and such — that call for extensive environmental assessment and management inside what the philosopher Peter Sloterdijk has provocatively called "the human zoo."[71] As the stakes of the protected area paradigm exceed the question of localized reserves to comprise a properly *global* environment, one within which humans too dwell, we can no longer imagine ourselves to stand over or outside such instrumental concerns. Instead, we find ourselves part of the biodiversity to be submitted to the technocratic expert rule. From a Heideggerian perspective, the Anthropocene thus figures as a symptom of a much more fundamental failure to pay attention to the manner in which being is revealed in any given historical and cultural setting, resulting in the unquestioned dominance of the technological way of revealing, which reduces the entire planet — including humans — to that of a resource in the service of an inherently meaningless project of incessant technological manipulation, a project whose name in its global-imperial form is "planetary stewardship."[72]

If we are to believe Heidegger, then, it is in the light of picturing humankind as the unrestricted master over its own destiny that we must understand the modern scientific desire to totalize and systematize the world in order to substitute God with an equally all-encompassing "theory of everything" — or, as Schellnhuber puts it with reference to the objective of earth system science, to develop "a rigorous common formalism, extracting the essence of all possible concepts."[73] To be sure, Heidegger's charge is against the anthropocentric hubris that leads humans to name a geological epoch after themselves — not because such an attitude holds the value of humankind too high, but because

71 I have borrowed the term "human zoo" from the English translation by Mary Varney Rorty; see Peter Sloterdijk, "*Rules for the Human Zoo:* A Response to the *Letter on Humanism*," trans. Mary V. Rorty, *Environment and Planning D: Society and Space* 27, no. 1 (2007): 12–28.

72 Steffen et al., "The Anthropocene."

73 Schellnhuber, "'Earth System' Analysis and the Second Copernican Revolution," 23.

it undervalues the human.[74] The Heideggerian rebuttal of the seemingly emancipatory consequences of the kind of humanism that replaces God with humankind is that it thereby also "completes subjectivity's unconditioned self-assertion."[75] Ironically, to embrace subjectivity unconditionally is to abandon freedom and self-determination in favor of the machine: purely subjective power has no values, no morality, no self-reflection, but is only the acting out of a blind will that takes its own extension as an end in itself.

If we consider this irony, we quickly begin to see why Heidegger was so unyielding in holding the view that his critique of humanism had better not eventuate in the existentialist sentiments of a post- or transhumanism.[76] By articulating a uniform

---

74 Martin Heidegger, "Letter on 'Humanism,'" in *Basic Writings,* ed. David F. Krell (New York: HarperCollins, 1978), 233–34.

75 Ibid., 244–45.

76 For a convincing argument as to the metaphysical convergence of post- and transhumanism onto an ontologically flat plane of immanence that disperses the subject in the name of instrumental reason, thereby bringing organicism and instrumentalism together in unholy marriage, see Luigi Pellizzoni, "New Materialism and Runaway Capitalism: A Critical Assessment," *Soft Power* 5, no. 1 (2017): 63–80. "An example," Pellizzoni points out, "comes from Rosi Braidotti's recent book on the post-human, the basic argument of which is that the 'dynamic, self-organizing, transversal force of life itself [...] conveyed by current technological transformations' — where life transmutes into technology and technology into life — is capable of 'displac[ing] the exploitative and necro-political gravitational pull of advanced capitalism.' This claim, it seems to me, fails to consider how such transformations are entrenched in runaway neoliberal capitalism, beginning with the type of subjectivity the former promotes and the latter presupposes (or vice versa): entrepreneurial, expansive, decentered, vitalistic" (Rosi Braidotti, quoted in ibid., 73–74).

In line with Heidegger's critique of humanism, most famously explicated in his "Letter on 'Humanism'" (1947), Pellizzoni holds that the post- and transhumanist objections both fall victim to the same metaphysical inversion upon whose grounds Heidegger denounced Jean-Paul Sartre's privileging of existence over essence. In other words, a proper overcoming of Western metaphysics cannot be as straightforward as a simple inversion of the Platonic eternal, rigid, and static being into contingent, dynamic, and processual becoming. Because, as Heidegger put it, "the reversal of a metaphysical statement remains a metaphysical statement" insofar as both

ontology of ecological homogeneity that maps beings onto a global network of matter-energy flow, viewing them either as resources to be harnessed or waste to be dispended, such a nihilistic attitude signals that our environment has become entirely technologized. Nature is no longer allowed to exist in its own right, intrinsically valuable, but appears solely in the instrumentalized sense of its use value — be it in terms of a carbon sink, faunal or entomological refuge, energy producer, and so on. Yet, far from disclosing the globalization of technology as a matter of planetary authoritarianism in the name of technological progression, whereby its imperial advance imposes "megamachines" of energy capture, resource extraction, and mass production[77] — its hydroelectric dams, its mechanized food industry, its assembly line factories, its agricultural irrigation systems, and such — upon a passive earth, the disclosure of planetary technicity not only instrumentalizes nature but concomitantly naturalizes technology. Urban environments come to be seen not as cultural artifacts that impose themselves upon an a priori geological and geographical location, but as hybrid designs that merely actualize geomorphological possibilities, and as ontologically inclusive assemblages that conjoin nature and artifice in the active production of terrestrial habitats.[78] To recast our nar-

sides to such a fundamental dualism are equally self-satisfied with having rendered that which is fully present to itself. See Heidegger, "Letter on 'Humanism,'" 232.

77 Lewis Mumford, *The Myth of the Machine,* vol. 1: *Technics and Human Development* (New York: Harcourt, 1967), 188–89.

78 Nigel Clark, *Inhuman Nature: Sociable Life on a Dynamic Planet* (London: Sage, 2011), 7–11, and Eva Lövbrand, Johannes Stripple, and Björn Wiman, "Earth System Governmentality: Reflections on Science in the Anthropocene," *Global Environmental Change* 19, no. 1 (2009): 11. For affirmative accounts of such hybridities, see Bruno Latour, "Love Your Monsters: Why We Must Care for Our Technologies as We Do Our Children," in *Love Your Monsters: Postenvironmentalism and the Anthropocene,* eds. Michael Schellenberger and Ted Nordhaus (Oakland: The Breakthrough Institute, 2011), 17–21; Bruno Latour, "A Cautious Prometheus? A Few Steps toward a Philosophy of Design with Special Attention to Peter Sloterdijk," in *In Medias Res: Peter Sloterdijk's Spherological Poetics of Being,* ed. Willem Schinkel and Liesbeth Noordegraaf-Eelens (Amsterdam: Amsterdam

rative frame on the basis of such a complete ontological reversal implies a movement from nature as the transcendental limit to human enterprise to nature as immanently produced by humankind. Hence, environmental change is ultimately inscribed into a neovitalist ontology of technical alteration that portrays the active modification and constant transgression of limits as the natural state of the geological economy. In this manner, the technification of nature appears as no less a desirable way of producing human existence than the conservationist ideal of letting nature itself dictate the terms of humanity's dwelling. Quite to the contrary, the former in fact gives the impression of being more desirable than the latter insofar as it maximizes the possibility for all products and forces to express themselves with maximum vitality by connecting them into a network such that all parts are allowed to increase their functional and expressive capacities in relation to each other.[79] The ensuing drama thus bridges the ontological abyss between humankind and other beings — an abyss that situates the human at a hermeneutic distance from the immediacy of the world, and whose bridging Heidegger sees as a threat to meditative thinking and in effect to human freedom.

§

University Press, 2011), 151–64; Erle C. Ellis, "The Planet of No Return," in *Love Your Monsters*, eds. Schellenberger and Nordhaus, 37–46; and Magdalena Hoły-Łuczaj and Vincent Blok, "How to Deal with Hybrids in the Anthropocene? Towards a Philosophy of Technology and Environmental Philosophy 2.0," *Environmental Values* 28, no. 3 (2019): 325–46.

79 Nigel Clark, "Rock, Life, Fire: Speculative Geophysics and the Anthropocene," *Oxford Literary Review* 34, no. 2 (2012): 259–76; Bronislaw Szerszynski, "The End of the End of Nature: The Anthropocene and the Fate of the Human," *Oxford Literary Review* 34, no. 2 (2012): 165–84; Bronislaw Szerszynski, "Planetary Mobilities: Movement, Memory, and Emergence in the Body of the Earth," *Mobilities* 11, no. 4 (2016): 614–28; and Bronislaw Szerszynski, "Out of the Metazoic? Animals as a Transitional Form in Planetary Evolution," in *Thinking about Animals in the Age of the Anthropocene*, eds. Morten Tønnesen, Kristin Armstrong Oma, and Silver Rattasepp (Lanham: Rowman & Littlefield, 2016), 163–79.

If, today, Heidegger's project of fundamental ontology and his investigation of the essence of technology in the context of the history of Western metaphysics seem hopelessly outdated, he can nevertheless be said to have articulated, in spite of the agony of his jargon, the question of our age: How is it that, at the historical moment in time when the power of technology has become recognized as an existential danger because of its capacity to alter the environment globally, technology has also come to be increasingly regarded as a natural process? For what has been fundamentally lacking, so far, in discussions about the Anthropocene, is a historical examination of the kind of ontological collapse between nature and artifice upon which earth system science has subsequently been able to base its production of knowledge.

First, although philosophers of technology seem to agree that the relevance of the Anthropocene primarily relates to a renewal of interest into questions concerning the essence of technology, much of the interest has nevertheless taken the historical circumstance of this ontological collapse at face value rather than entertaining the Kantian question about the preconditions that made such a disclosure possible in the first place. In *The Neganthropocene* (2018), for instance, the French philosopher Bernard Stiegler has framed the question concerning technology in a narrative that emphasizes the toxic wastelands of the Anthropocene as an era of entropic decay and waste. For Stiegler, the Anthropocene marks humanity's coming to terms with the fact that it now dwells within an inherently unsustainable global-technological system that can maintain itself only by accelerating its own demise, through ever-increasing efficiency in producing what in thermodynamic terms is known as "entropy." This is a mode of production guided by a quasi-teleological attractor that draws the earth's history over a series of intensive, qualitative thresholds that have no eschatological point of completion, reaching self-termination only when its substrate of resources has been entirely exhausted. It is assumed, however, that our era is fundamentally debt-driven, bound to a conception of technology as an automated entropic machine of accumulation and dissipation. Although Stiegler's thesis is that the entropic nature of our current mode of production can — or rather, must — be overcome by realigning technology with the negentropic nature of life, the question of what makes the former conception an obvious starting point for addressing the problem of technology in the face of global environmental change, in opposition to which the second conception then appears as its given solution, is not itself warranted any further examination.

In much the same way, scholars writing on the implications of the Anthropocene for the ontology of technology have arguably failed to navigate a middle way between a nostalgic return to metaphysics, on the one hand, and a no-nonsense antiessentialism, on the other. At one end, the ontological collapse between nature and artifice has been dogmatically affirmed without recourse to the conditions under which it has been disclosed as such; at the other end, the boundaries of the technological as a separate domain have been completely taken apart in favor of the immanent production of a self-organizing earth system. From this point of view, the smooth and networked being of entities has been unreservedly accepted as given, such that any effort to critically limit the technological has been a priori deemed an illegitimate restriction of its scope, since it would in effect neglect the terrestrial environment within which technical artifacts and practices always and already operate. The reflexive response among philosophers of technology to this supposedly new geological condition has thus been to uncritically "root" technology in its planetary milieu,[80] in effect making philosophy a handmaid to the sciences by abandoning its hermeneutic task. In short, interesting work in the philosophy of technology has admittedly been produced in an effort to understand the implications of the Anthropocene as a geological epoch, but far too little attention has been paid to the historical circumstances of the Anthropocene as a technical term, that is, as the object of

80 Vincent Blok, "Earthing Technology: Towards an Eco-Centric Concept of Biomimetic Technologies in the Anthropocene," *Techné: Research in Philosophy and Technology* 21, nos. 2–3 (2017): 127–49.

a scientific paradigm. This is not to argue that the latter should hold primacy over the former, but merely to draw attention to the fact that the dominance of the former over the latter risks eroding the critical potential of philosophy altogether.

Second, if philosophers of technology have inadequately mobilized the historical resources of their discipline, then historians of science and technology, for their part, have paid far from sufficient attention to the normative discursive horizon of the Anthropocene. For instance, in their effort to complement a survey of the development of modern climate science with an economic history of industrial society, historians Christophe Bonneuil and Jean-Baptiste Fressoz guide the reader through the various terms and points of contention in the Anthropocene discourse in order to carefully examine them within an ontologically unified context that calls for a reunion of human history with natural history—in effect positioning their work in the vein of Marxist scholarship that operates on undermining the ontological distinction between nature and artifice, typified by the work of Jason W. Moore.[81] "In the Anthropocene," Bonneuil and Fressoz write, "it is impossible to hide *the fact* that 'social' relations are full of biophysical processes, and that the various flows of matter and energy that run through the Earth system at different levels are polarized by socially structured human activities."[82] Since humans no longer act against the backdrop of an unchangeable nature, their technical enterprise

81 Jason W. Moore, "Transcending the Metabolic Rift: A Theory of Crises in the Capitalist World-Ecology," *Journal of Peasant Studies* 38, no. 1 (2011): 1–46; "Toward a Singular Metabolism: Epistemic Rifts and Environment-Making in the Capitalist World-Ecology," in *New Geographies*, 6: *Grounding Metabolism*, eds. Daniel Ibañez and Nikos Katsikis (Cambridge: Harvard University Press, 2014), 10–19; *Capitalism in the Web of Life: Ecology and the Accumulation of Capital* (London: Verso, 2015); and "Anthropocene or Capitalocene? Nature, History, and the Crisis of Capitalism," in *Anthropocene or Capitalocene? Nature, History, and the Crisis of Capitalism*, ed. Jason W. Moore (Oakland: Kairos, 2016), 1–11.

82 Christophe Bonneuil and Jean-Baptiste Fressoz, *The Shock of the Anthropocene: The Earth, History and Us* (London: Verso, 2016), 39 (my emphasis).

instead being deeply woven into its very fabric, this new condition supposedly needs to be historicized not only as a history of the environment but as a history of technology too. As one of the leading proponents of this "terrestrial turn," Jürgen Renn, has put it, "the *new reality* of the planet confronts us with a [...] radical need for rethinking our situation: we are not living in a stable environment that simply serves as a stage and resource for our actions, but we are all actors in a comprehensive drama in which humans and the nonhuman world equally take part."[83] In the same vein, the historians of technology Sverker Sörlin and Nina Wormbs have suggested that we ought to understand the essence of technology, in the face of the Anthropocene, as a "practice of terraforming."[84] They argue that modern historiography has undergone such a profound reorganization through the destabilization and rearticulation of its binary mode of categorization—which has served as the basis for understanding the relationship between humanity and nature ever since the inception of history as an academic discipline—that the category neither of nature nor of artifice allows for sufficient conceptual work as long as each is kept dualistically apart from the other. Their contention is that in order to come to terms with a planet characterized by natural and artificial processes of change that dynamically interact to reproduce metastable conditions for life, and in the midst of which humankind's remarkable ability to modify nature can no longer be seen as a curious exception but rather as integral to the workings of the geological economy, it is necessary to rethink technology beyond the nature/artifice dichotomy.

The same lack of critical distance toward the horizon of understanding underlying the sciences can be discerned in the world historian William H. McNeill's call for an intellectual partnership between natural scientists and humanities schol-

83 Jürgen Renn, "The Evolution of Knowledge: Rethinking Science in the Anthropocene," *Journal of History of Science and Technology* 12, no. 1 (2018): 3 (my emphasis).

84 Sverker Sörlin and Nina Wormbs, "Environing Technologies: A Theory of Making Environment," *History and Technology* 34, no. 2 (2018): 101–25.

ars, arguing already in 2001 that "it is time for historians to [...] begin to connect their own professional thinking and writing with *the revised scientific version of the nature of things.*"[85] Yet, the unspoken terms — that is, the epistemic and ontological tenets of complexity science and managerial systems thinking — upon which we ought to "come to terms" with this particular understanding of the earth have themselves tended to remain critically unexamined and have thus also been left largely outside the framework within which the Anthropocene has been historicized. Curiously omitted is the complexity-parlance itself, within whose confines we remain prisoners to a particular historical horizon. In other words, the danger with the post-Enlightenment myth of progress is when it becomes difficult to separate knowledge from the conditions of its establishment, such that the historicization of the Anthropocene, and thus the historicization of the unity of nature and artifice in the earth system, still proceeds from the same metaphysical assumptions as its object of study. There is still the sense, then, that even if the emergence of the concern for humanity as a geological agent has already been meticulously situated in the context of post-war systems science, such a historicization remains incomplete as long as it deals exclusively with empirics at the expense of the transcendental conditions for its disclosure, and especially when it does so to such a degree that its present disclosure may uncritically appear as universal and self-evident.

For these very reasons, the potential for an intellectually stimulating interaction between philosophy and history has, in the case of the question concerning planetary technicity, been stunted by the institutionalization of two largely isolated approaches that have only tangentially met in the middle. More specifically, though numerous scholars have already sought to respond to the implications of the Anthropocene for the general philosophical understanding of technology, on the one hand,

85 William H. McNeill, "Passing Strange: The Convergence of Evolutionary Science with Scientific History," *History and Theory* 40, no. 1 (2001): 5 (my emphasis).

and to elucidate the context of knowledge production behind the Anthropocene, on the other, there has been a surprising lack of effort in bringing these two projects together,[86] that is, bringing a philosophical investigation of the nature/artifice binary into conversation with an intellectual historical investigation of the presence of such ontological concerns within the earth sciences. One such productive but so far also underdeveloped exception has been proposed by a number of scholars who have pursued an interest in critically investigating the ideological aspect of the Anthropocene.[87] For instance, in his review of the concept, Jeremy Baskin has shifted attention to the pairing of descriptive and prescriptive dimensions and has demonstrated how the framing of the problem of global environmental change presupposes the possibility of assimilating technological processes into the natural order of things. According to Baskin, proponents of this "naturalization" of technology include Crutzen, according to whom "we should shift our mission from crusade to management, so we can steer nature's course *symbiotically*," and the geographer Erle Ellis, who states that "in moving toward a better

86 It is especially surprising considering that many of the ideas that have been identified as important historical precursors to the Anthropocene have been products of the attempt to rethink humankind's relationship to nature and, in effect, to rethink technology as the mediator of this interaction. For even though awareness of this dynamic may seem incredibly contemporary, it actually dates surprisingly far back into history. As pointed out in an article coauthored by Crutzen, this is far from the first time that humans have attested to or foreseen such technological power over the fate of the planet — whether to celebrate it or as a cause for concern. See Will Steffen et al., "The Anthropocene: Conceptual and Historical Perspectives," *Philosophical Transactions of the Royal Society A* 369, no. 1938 (2011): 842–67.

87 For a convincing call for more research of this kind, see Eva Lövbrand et al., "Who Speaks for the Future of Earth? How Critical Social Science Can Extend the Conversation on the Anthropocene," *Global Environmental Change* 32 (2015): 211–18. For a call for the genealogical investigation of the ontological flattening of the artificial vis-à-vis the natural in earth system science, see Lövbrand, Stripple, and Wiman, "Earth System Governmentality," and Rolf Lidskog and Claire Waterton, "Anthropocene — A Cautious Welcome from Environmental Sociology?," *Environmental Sociology* 2, no. 4 (2016): 402–3.

Anthropocene, the environment will be what we make it."[88] Certainly, there are differences in tone among the various accounts, as demonstrated by these two quotes. On the one hand, there are those for whom nature never was truly "natural" in the first place, and who enthusiastically embrace the Anthropocene condition for finally giving technology free rein to work the earth according to nothing but the whim of the free market, and then, on the other hand, there are those for whom technology never was "artificial" or "unnatural," but since the industrial evolution has become "out of sync" with its terrestrial environment, such that the Anthropocene for them represents the need to bridge the gap between nature and artifice again. What both accounts share, however, is an implicit ontological breakdown of the dichotomy between nature and artifice into a monistic hybrid, whereby it is not only possible but in fact desirable for humans to actively participate in the betterment of nature.

Pursuing the same line of argument as Baskin, but focusing exclusively on technology, the Italian philosopher Agostino Cera holds that this ideological vein runs to the very core of the Anthropocene discourse, since, as he writes,

> quite unquestioningly, it expresses the accepted meaning of an epochal fact, i.e., the complete and definitive *naturalization of technology.* The normative/prescriptive element of this aspirant geological epoch lies in its unquestioning, "natural" acceptance of the *metamorphosis of technē in phusis.* In other words: within the present-day historical configuration, technology has taken on such a pervasive role that the only way it can be properly perceived is to think of it and interpret it as being nature itself.[89]

---

88 Paul J. Crutzen and Erle C. Ellis, quoted in Jeremy Baskin, "Paradigm Dressed as Epoch: The Ideology of the Anthropocene," *Environmental Values* 24 (2015): 14, 17 (my emphasis).

89 Agostino Cera, "The Technocene or Technology as (Neo)Environment," *Techné: Research in Philosophy and Technology* 21, nos. 2–3 (2017): 247.

As in the case with Baskin's critique of ideology, Cera argues in a similar manner that the Anthropocene is conditioned by an implicit integration of technology into the self-coinciding identity of the earth as a systematic whole—which, however, can appear as self-coinciding only insofar as its own conditions of possibility remain unexamined. In short, such a concern has fundamentally to do with the elimination of an object-disclosing horizon altogether, which, by emancipating our discourse from the conditions for epistemic access to the world simultaneously forms an ontology liberated from any demand for justification and argumentation. As opposed to a *critical* theory, which is wedded to conceive dialectically of the interpenetration of nature and artifice, the latter presupposes a fundamental undecidability between the two and thus a priori dissolves the need for this kind of critical work. For if the underlying ontological collapse of the distinction between the natural and the artificial in the Anthropocene has come to be increasingly accepted as self-evident, then it is necessary to examine the naturalization of technology in the context of its wider discursive formation, precisely in order to avoid the superficial equation of the laws of nature with the commodifying logic of capitalism. For what has been severely lacking when it comes to the question concerning the essence of technology lately is precisely a philosophical historical critique of the present: a bringing-to-the-fore of the presuppositions behind its epochal disclosure as indistinguishable from nature. Without such a critique, as Baskin notes, we run a serious risk of unreflexively adopting ideas such as Ellis's, namely, that nature is nothing other than "what we make it," nothing other than its commodification.

§

This should be reason enough to turn to the genealogical tradition, running from Friedrich Nietzsche's genealogy of morals through Heidegger's *Seinsgeschichte* up to Michel Foucault's genealogy of power. Only with the help of the genealogical toolbox can we examine the dangers to which we are blind because they come before, or underpin, our problems such that we can begin to outline and devise currently "unthinkable" alternatives. Put concisely by the historian Leo Marx:

> Understanding these changes is complicated [...] by the fact that the most fitting language for describing them came into being as a result of, and indeed largely in response to, these very changes. The crucial case is that of "technology" itself. To be sure, we intuitively account for the currency of the word in its broad modern sense as an obvious reflex of the increasing proliferation of [...] new and more powerful machinery. But, again, that truism is not an adequate historical explanation. It reveals nothing about the preconditions—the specific conceptual or expressive needs unsatisfied by the previously existing vocabulary[. ...] Such an inquiry is not trivial, nor is it merely semantic.[90]

In this sense, we might surmise that there is more to the phenomenon of globalization than a transformation of the practices of organization, management, and governance through market-oriented arrangements. As the British geographer Stuart Elden has argued, the spatial extension implied in the concept of globalization hinges on an ontological conception of the casting of space. We therefore also need to attend to questions that are "not concerned with 'what is,' but with *how* 'what is' *is*."[91] To examine planetary technicity as a mode of disclosure means contemplating how global spatiality corresponds to a certain historically conditioned understanding of humankind's being on the earth. As such, Heidegger's paradoxical-sounding claim that the essence of technology is nothing technological does not mean that technology today is so all-encompassing that there

90 Leo Marx, "The Idea of 'Technology' and Postmodern Pessimism," in *Does Technology Drive History? The Dilemma of Technological Determinism*, eds. Merritt R. Smith and Leo Marx (Cambridge: MIT Press, 1994), 241–42.

91 Stuart Elden, "Missing the Point: Globalization, Deterritorialization, and the Space of the World," *Transactions of the Institute of British Geographers* 30, no. 1 (2005): 16.

is no room for reflexivity, but precisely the opposite: it is only by attending to its mode of disclosure that we can begin to discern the limits to our current horizon of understanding.[92] The environmental sociologist Luigi Pellizzoni writes that it is "only in this way — that is, taking a genealogical outlook — [that] one may go beyond questions of intellectual fashions or academic disputes [to grasp] the actual stakes of the issue, namely, how a burgeoning governmentality builds on what would purportedly offer the basis for an effective critique."[93] We must therefore ask, What is the necessary framework for the appearance of our present technological condition in terms of self-organizing global networks?

The methodological premise of this book thus takes its cue from Imre Lakatos's well-known paraphrase of Immanuel Kant's maxim: "Philosophy of science without history of science is empty; history of science without philosophy of science is blind."[94] The same maxim, I believe, applies equally well to the study of technology. For the more general point to take away from Lakatos's observation is that whenever history and philosophy intermingle, we are required to put aside both the metaphysical cravings that so often occlude the vision of philosophers, whether characterized by the desire to uncover a moment of original purity or that of a universal a priori framework, and the Whiggish notion of a linear history from past to present, which, on its own, lacks a critical distance between history and the horizon for its rational reconstruction. In other words, in the fruitful marriage between history and philosophy we are forced to recognize not only that our knowledge and evidence changes throughout history, but also that our conceptual understanding of what constitutes, for instance, "artifice" as opposed to "nature,"

92 Iain D. Thomson, "From the Question Concerning Technology to the Quest for a Democratic Technology: Heidegger, Marcuse, Feenberg," *Inquiry: An Interdisciplinary Journal of Philosophy* 43, no. 2 (2000): 203–16.

93 Pellizzoni, "New Materialism and Runaway Capitalism," 65.

94 Imre Lakatos, "History of Science and Its Rational Reconstructions," in *Scientific Revolutions*, ed. Ian Hacking (Oxford: Oxford University Press, 1981), 107.

can be historicized. After all, history cannot be reduced to the accumulation of facts, for it is as much an interrogative and corrective discourse of active revisioning, having at least as much to do with the discursive structure within which facts become meaningful. This book is born out of the conviction that a historiography attuned simultaneously to theories and hypotheses, the currents of ideas within which they develop, and to their ontological conjectures, has a precise function: it allows us to recognize a real and useful connection between history and philosophy. This connection is important for two reasons. First, to recontextualize traditional philosophical questions about the essence of technology in light of our contemporary social and political concerns of the Anthropocene, and, second, as a means of assessing our present beliefs and means of talking about technology in light of rich philosophical reflections on the topic. This book in effect seeks, on the one hand, to initiate a discussion about the ontology of technology relevant for the sake of a cultural critique of our present situation, and, on the other, to attach the discussion about the role of technology in the face of global environmental change to philosophical considerations of its meaning.

Of course, recent scholarship in the philosophy of technology has generally eschewed ontological questions altogether, preferring instead to concentrate on constructivist models or pragmatic analyses. In this book, however, I intend to return to hermeneutics, seeking not so much a definitive essence of technology as a historical analysis of the conditions for its disclosure as indistinguishable from nature. Seen from this perspective, the key issue is not how humanity can or should realize itself in the world with the help of technology, but how and in what way the world appears to humans through various mode of disclosure. Such modes are not solely of academic or intellectual interest but have very practical consequences in the manner that they shape the possibilities for knowledge, belief, and ultimately action within determinate historical circumstances. Consequently, this book is positioned on an ontological rather than ontic level, because it is concerned with the nature of specific kinds of objects rather than what an object can or does do in the phenomenal realm. And yet it is also historical in that it does not content itself with interpreting historically specific phenomena in terms of a transhistorical essence, but rather seeks to understand the conditions for the particular horizon of understanding within which phenomena are disclosed. In this sense, the historicity of being serves as a valuable point of departure for us inasmuch as it shifts the focus of the question concerning technology from the constricted and fixed confines of pure reason to the genealogically rich field of meditations on its essence that inhabit intellectual history and that demonstrate, through their historical concreteness, the historical condition of philosophical meditation. In order to fully grasp the stakes of its discourse, we need not only understand the contemporary scientific and technological transformations behind the Anthropocene, but also explore the history of an ontological concern tied up with it. To follow this philosophical trajectory is to learn how we arrived at this critical moment in history, and to know where we might head in the twenty-first century.

2

Deep Time of the Heat Engine

Although recent discussions centering on the idea of an earth system grew out of transdisciplinary efforts in the late twentieth century, recognition that the interaction between geological, chemical, biological, and even technological processes of change may be significant goes back almost two and a half centuries to at least the geological musings of the eighteenth-century Scottish naturalist and physician James Hutton. A major figure in the Scottish Enlightenment, Hutton had already begun in 1785 to make public the first sketches of a geotheory that he had been working on over the past two decades. As the historian Martin Rudwick has demonstrated, geotheory was a genre that sought to emulate the hypothetico-deductive approach by which Isaac Newton had advanced his theory of the solar system, and to apply it to geology in an effort to develop a comprehensive model that would similarly provide a systematic explanation of geophysical processes. Consequently, such a general theory was judged to be the crowning achievement of an eighteenth-century naturalist. Natural historians and philosophers alike would spend the majority of their careers carrying out the more mundane work of reviewing literature and gathering evidence in the hope that it would allow them to eventually present a grand synthesis founded upon a few comprehensive laws, with the expectation that these laws would illuminate the mechanisms of humankind's terrestrial home much in the same way as Newton had provided an explanation of its cosmic abode[1] — what the media scholar Siegfried Zielinski, borrowing the term from Novalis's *Heinrich von Ofterdingen* (1802), calls an "inverted astronomy."[2] In other words, overarching theories of the earth and its history were constructed to explain and rationalize all known facts. Although they certainly sought to incorporate empirical findings, geotheories seamlessly bridged empirics with a kind of speculation that was ultimately unbridled by observational rigor.

By the turn of the century, geotheory had already fallen out of favor in the English-speaking world to be replaced by inductive investigation. Among the British naturalists, an empirical view of scientific knowledge quickly came to dominate, principally because of the Scottish Enlightenment's most prominent figure, David Hume, who had persuasively rendered invalid any scientific explanation referring to an ultimate cause whose existence could not itself be the subject of observation. Most famously laid out in his *Enquiries Concerning Human Understanding* (1777), Hume noted therein that "as to past *Experience*, it can be allowed to give *direct* and *certain* information of those precise objects only, and that precise period of time, which fell under its cognizance."[3] Historians largely agree that nineteenth-century geology therefore came to eschew theoretical conjecture in favor of rigorous empirical studies based upon observable causes.[4] In fact, the formation of the Geological Society of London in 1807 was deliberately planned as a corrective to those

1 Martin J.S. Rudwick, *Bursting the Limits of Time: The Reconstruction of Geohistory in the Age of Revolution* (Chicago: University of Chicago Press, 2005), 133–39.

2 Siegfried Zielinski, *Deep Time of the Media: Toward an Archaeology of Hearing and Seeing by Technical Means*, trans. Gloria Custance (Cambridge: MIT Press, 2006), 18–25.

3 David Hume, *Enquiries Concerning Human Understanding and Concerning the Principles of Morals* (Oxford: Clarendon Press, 1902), 3.

4 Rudwick, *Bursting the Limits of Time*; James A. Secord, *Controversy in Victorian Geology: The Cambrian–Silurian Dispute* (Princeton: Princeton University Press, 2014); and David R. Oldroyd, *The Highlands Controversy:*

speculative cosmologies that had previously laid claim to grand narratives of the earth, disparaging the effort to imaginatively re-create the past insofar as it leads into the landscape of fiction. In their loathing of theory, the founding members of the Geological Society were suspicious of any reference to narrative with its emphasis on motive. This methodological premise quickly became integral to the practice of geology.[5] Further cemented by the posthumous popularization of Hutton's work by Charles Lyell, and the coinage of the term "uniformitarianism" by William Whewell to describe the methodological implications of his eternalism, geology as a discipline came to be characterized precisely by its hostility toward geogony and toward concerns "'with questions as to the origin of things.'"[6] Hence, by the 1830s, geologists were far more concerned with the order and structure of strata than they were with reconstructing the earth's history and genesis.

But then again, this was not yet the case when Hutton was writing, nor was such a constrained approach compatible with his own concerns. Although he stood on the threshold of a new age, writing during a time in which geology was in the midst of its institutionalization as a scientific discipline, slowly being grounded in empirical observation and thus freed from the restrictions of theology, as an eighteenth-century naturalist Hutton was no foreigner to arguing from an ontotheological standpoint. Although God played a minor role, it was still common during the eighteenth century to demonstrate how God's power, wisdom, and goodness manifested itself in creation. Before natural science had been emancipated from its Christian

Constructing Geological Knowledge through Fieldwork in Nineteenth-Century Britain (Chicago: University of Chicago Press, 1990).

5 Martin J.S. Rudwick, *Earth's Deep History: How It Was Discovered and Why It Matters* (Chicago: University of Chicago Press, 2014), 156–57, and Martin J.S. Rudwick, *The Great Devonian Controversy: The Shaping of Scientific Knowledge among Gentlemanly Specialists* (Chicago: University of Chicago Press, 1985), 63–68.

6 Charles Lyell, *Principles of Geology: Being an Inquiry How Far the Former Changes of the Earth's Surface Are Referable to Causes Now in Operation*, Vol. 1 (London: John Murray, 1835), 5.

framing, natural theology—or, at least a vague reference to the Creator's omnipotence and wisdom—was a matter of course for the naturalist. In their work, theology remained alive, albeit in a more worldly form. Certainly, it was claimed that science provided evidence of the existence of a higher order, but the focus was on nature as such rather than on speculations that pertained to the Creator. In fact, God was made so abstract as to become just another word for the uniformity of nature. And with the rise of modern science and the increasing confidence in the ability of human reason to penetrate the mysteries of nature, natural theology came to strengthen its position. Nature was ordained by God and preserved in the form it had originally received, and consequently natural science should not try to explain how things had arisen but to name, classify, and describe what currently existed as a comprehensive natural system.[7] Although Copernican cosmology, Newtonian mechanics, and Cartesian substance dualism challenged the seemingly eternal, logical, and inherently rational Aristotelean worldview, it was not an easily shaken system that these early moderns were confronted with. The geosomatic depiction of the earth as a kind of superorganism—popular during Greek antiquity—never fully recovered from the onslaught of the scientific revolution, but it would, as we shall see, nevertheless return in modernity in an inverted form.

Geology beyond Mechanism

Before the turn of the nineteenth century, the geoscientific landscape was remarkably different than compared to only a few years later. In the late eighteenth century, when Hutton was at the pinnacle of his career, the principal geological controversy was the dispute between the "Neptunists" and their adversaries the "Plutonists." Whereas the Neptunists—aptly denominated after the Roman deity of water—held that catastrophic geologi-

7 Rudwick, *Earth's Deep History*, 161, and Dennis R. Dean, *James Hutton and the History of Geology* (Ithaca: Cornell University Press, 1992), 2, 5–6.

cal changes in the form of huge floods had shaped what was observed at the time, the Plutonists — a designation that stems from the ruler of the underworld in Greco-Roman mythology — argued, on the contrary, that the main agency was magmatic activity, which, rather than driving geological evolution through sudden events, slowly played out across unimaginable time spans. At the time, many of the savants of the Western world were still clinging to the Neptunian theory of the earth proposed by the German mineralogist Abraham Gottlob Werner, whose distinguishing postulation was that the solid landforms of our planet had been fashioned, once and for all, beneath a primeval ocean, and from which it followed that most rocks observed in the strata — apart from the effects of the occasional eruption of magma — were sedimentary deposits whose sequence, when read, would reveal their historical formation in an orderly manner.[8] But Hutton, with the presentation of his geotheory, came to challenge the Neptunist model in several fundamental ways. When investigating the rocks of his native Scotland, Hutton observed fingers of granite reaching well into the sedimentary rocks — an observation that seemed to contradict the neat stratification predicted by the Neptunists, and that he believed to point toward the formative character of the element classically understood as diametrically opposite to water: subterranean fire and heat.[9] In three successive summers, Hutton found veins of granite penetrating the schist, first in the Highlands, then in the hills of Galloway in the southern Uplands, and then on the Isle of Arran off the west coast.[10] Such veins had already been described, but Hutton interpreted their implications differently, as evidence that the granite had been squirted into the other rock from below, as a hot fluid that had crystallized as it

8 Rudwick, *Bursting the Limits of Time*, 421–23.

9 James Hutton, *Theory of the Earth, with Proofs and Illustrations*, Vol. 1 (Edinburgh: William Creech, 1795), 317–18.

10 John Playfair, "Biographical Account of the Late Dr. James Hutton," *Transactions of the Royal Society of Edinburgh* 5 (1805): 68–69, and James Hutton, "Observations on Granite," *Transactions of the Royal Society of Edinburgh* 3 (1794): 79–80.

cooled. This implied that the lowest rock mass in the geognostic pile was not in fact foundational, so that there might not be any truly primitive rocks at all.[11] Furthermore, if granites were introduced from below, they would become strong evidence of great heat, and of the required agent of elevation — the analogue of steam in the steam engine, forcing the crust upward to form a new landmass. Similar evidence for the forcible elevation of landmass came from Hutton's interpretation of the many layers of whinstone or basalt intercalated among the other secondary formations. He thus adopted the conclusion that basalt was a rock of volcanic origin, but added that it was material that had been forcefully inserted into the pile of sediments deep within the earth — forming, in geological terms, an intrusive sill — thereby contributing to crustal elevation without reaching the surface as lava in a volcanic eruption. Indeed, he argued that volcanoes were simply nature's safety valves, regulating and preventing excessive pressure from below. Hence, volcanoes were, from Hutton's point of view, a means by which a designful order was dynamically maintained as opposed to an already finished product.

In fact, Hutton likened the earth precisely to a "beautiful machine," which, just like machines of human origin, was artfully designed and constructed in order to achieve an intended outcome: "When we trace the parts of which this terrestrial system is composed, and when we view the general connection of those several parts, the whole presents a machine of a peculiar construction by which it is adapted to a certain end. We perceive a fabric, erected in wisdom, to obtain a purpose worthy of the power that is apparent in the production of it."[12] It is important to remember that Hutton did not live, as we do, surrounded by a bewildering variety of machines. He and his contemporaries understood by that word one specific device above all others:

11 Hutton, *Theory of the Earth,* 1:311–12.

12 James Hutton, "Theory of the Earth; or an Investigation of the Laws Observable in the Composition, Dissolution, and Restoration of Land upon the Globe," *Transactions of the Royal Society of Edinburgh* 1, no. 2 (1788): 209.

the steam engine. Steam engines dominated the new industrial scene at the time. As the philosopher Michel Serres has stressed, it is no coincidence that for those at the heart of the industrial revolution, nature itself began to appear in the guise of the blazing and energetic transformations occurring inside these mechanical heat engines, unconsciously revealing the uneasiness of a social order being transformed by fiery energy, as a premodern way of life pushed along by wind, water flow, and muscle swiftly submitted to a modern world propelled by steam.[13] Although the improved steam engine devised by Hutton's Edinburgh contemporary James Watt was still a novelty in the late 1780s, the earlier, slower, and cruder Newcomen engine was in fact a more apt analogy for what Hutton had in mind.[14] The rise of the Newcomen engine's beam by the expansion of steam was a highly appropriate analogy for his notion of crustal elevation. The sheer irresistible power of steam was what impressed all who witnessed these engines in operation, and it made the machine an equally powerful image to convey Hutton's argument for the steady state of the earth, based on huge unforeseen forces deep below the surface. Just how those forces worked in the depth of the earth was what he tried to elucidate through his physics of heat—a major topic in his other writings. But in any case, it was clear that heat represented an expansive force that was in perpetual interaction with its opposite, the contractive force of gravitation. The oscillation of the Newcomen engine was an eloquent image of the same process of uplift and downthrust in nature. It was also a perfect metaphor in the sense that it implied the existence of a divine artificer—however unnecessary a hypothesis, as Pierre-Simon de Laplace is said to have put it in his apocryphal interaction

13 Michel Serres, "Turner Translates Carnot," in *Hermes: Literature, Science, Philosophy*, eds. Josué V. Harari and David F. Bell (Baltimore: Johns Hopkins University Press, 1982), 58–59. See also Nigel Clark, "Earth, Fire, Art: Pyrotechnology and the Crafting of the Social," in *Inventing the Social*, eds. Noortje Marres, Michael Guggenheim, and Alex Wilkie (Manchester: Mattering Press, 2018), 173–74.

14 Rudwick, *Bursting the Limits of Time*, 161.

with Napoleon Bonaparte — that had purposefully designed the planetary heat engine.

Armed with his machine metaphor, Hutton went about challenging the popularity of Werner's geotheory. In the first volume of his *Theory of the Earth, with Proofs and Illustrations* (1795), he recalled that "one day, walking in the beautiful valley above the town of Jedburgh, I was surprised with the appearance of vertical strata in the bed of the river, where I was certain that the banks were composed of horizontal strata. I was soon satisfied with regard to this phenomenon, and rejoiced at my good fortune in stumbling upon an object so interesting to the natural history of the earth, and which I had been long looking for in vain."[15] Of particular interest to Hutton, though, was his observation that "above those vertical strata, [were] placed the horizontal beds, which extend along the whole country,"[16] an insight that caused him to spend a substantial amount of time considering how such a juxtaposition could have come about and what the implications were for the explanatory power of Neptunism. Well acquainted with the state of the field, Hutton knew that Neptunists would argue that the schist had been shaped from sediment that had accumulated at the seafloor. In contrast, he suggested that that these strata had in fact been hardened by subterraneous heat and pressure, then folded into an upright orientation by these same forces, and consequently, over deep time, generated regions of elevated terrain by raising them well above the surface of the ocean. But because of the combined effects of water and wind slowly exposing and wearing down the rock, the making of new landmasses were at the same time countered by erosive forces upon the old, such that "this surface [would sink] below the influence of those destructive operations, and thus placed in a situation proper for the opposite effect, the accumulation of matter prepared and put in motion by the destroying causes,"[17] thus reprocessing its materials to form an essentially

15 Hutton, *Theory of the Earth,* 1:432.
16 Ibid.
17 Ibid., 1:435.

cyclical system of deterioration and regeneration. As Hutton pointed out, the angular unconformity in question further complicated the Neptunist theory by the fact that he found puddingstone interposed between the horizontal and vertical strata, whose hardened state he interpreted as evidence that the eroded schist had once more been subject to the hardening effects of heat and pressure. Witnessing neatly deposited strata of sedimentary rock overlaying almost vertical layers, he concluded that the lower levels must have been deposited eons before, but later been upturned. From Hutton's point of view, the horizontal beds of Old Red Sandstone at Siccar Point near Jedburgh must thus have been laid down upon the indurated puddingstone at the bottom of the ocean and subsequently consolidated by Plutonic forces, in whose wake the entire structure had been raised above sea level, only to be laid bare by erosive effects.[18]

As a consequence, Hutton deduced that the core of the earth was continually reproducing hardened rock to offset the effects of erosion. It must be, he surmised, that intrusive magmatic activity is implicated in the construction of elevated terrain, and that the general stirring of subterraneous forces gradually turns the seabed into a mountaintop.[19] Plutonism, Hutton maintained, is *the* great power in the global dynamic of the earth, in the sense that it is absolutely essential for its constitution as an enduring whole. As he saw it, the primary agency in the history of the earth was thus not oceanic precipitation, but magmatic intrusions and eruptions, such that the observed landmasses had not once and for all emerged out of the crystallization and sedimentation of rocks at the bottom of a universal ocean, but had rather been raised—and were continually being raised, albeit at a rate imperceptible to the human senses—by the volatile forces of the underworld. This was a radically creative destructive force: once the orderly landmasses had elevated above water

18 Tom Furniss, "James Hutton's Geological Tours of Scotland: Romanticism, Literary Strategies, and the Scientific Quest," *Science & Education* 23, no. 3 (2014): 565–88.

19 Hutton, *Theory of the Earth,* 1:330–31.

level, they were immediately subject to forces of erosion, yet, without erosion, there would be no terrestrial life on the planet, such as plants and animals: "A solid body of land could not have answered the purpose of a habitable world; for a soil is necessary to the growth of plants; and a soil is nothing but the materials collected from the destruction of the solid land. Therefore, the surface of this land, inhabited by man, and covered with plants and animals, is made by nature to decay, in dissolving from that hard and compact state in which it is found below the soil."[20]

But he noted that once the soil had been fashioned out of its previously solid state, it too "is necessarily washed away, by the continual circulation of the water, running from the summits of the mountains towards the general receptacle of that fluid," so that eventually, "by the agitation of the winds, the tides and currents, every moveable thing is carried farther and farther along the shelving bottom of the sea, towards the unfathomable regions of the ocean."[21] For this reason, Hutton argued, "we are not to look for nature in a quiescent state; matter itself must be in motion, and the scenes of life a continued or repeated series of agitations and events."[22] Neptunist theories held that the oceanic formation of the earth's geomorphology had taken place once, either in the recent or distant past, so as to imply that the planet was first and foremost a static and stable being only occasionally perturbed by out-of-the-ordinary events, but Hutton instead turned this notion on its head to argue that constant movement was in fact its natural state — landforms were incessantly being uplifted, but since these creative process were kept in check by their dynamic negotiation with an equal part of destruction, the planet as a whole would potentially oscillate between different stable states but nevertheless reproduce itself in perpetuity.

Thus, Hutton's machine analogy would turn out to be insufficient for describing the constitution of the earth insofar as mechanism implied mere linear cause and effect. "It is not only

20 Hutton, "Theory of the Earth," 214.
21 Ibid., 214–15.
22 Ibid., 209.

by seeing those general operations of the globe which depend upon its peculiar contribution as a machine," he wrote, "but also by perceiving how far the particular, in the construction of that machine, depend upon the general operations of the globe, that we are enabled to understand the constitution of this earth."[23] Without completely abandoning it, Hutton therefore complemented his machine-metaphor with that of an organism. As he rhetorically put it: "But is this world to be considered thus merely a machine to last no longer than its parts retain their present position, their proper forms and qualities? Or may it not be also considered an organised body? Such as has a constitution in which the necessary decay of the machine is naturally repaired, in the exertion of those productive powers by which it had been formed."[24]

For Hutton, the earth's incessant renewal of its eroding topography recalled the same process of growth and repair that restored the organism, in that "this earth, like the body of an animal, is wasted at the same time as it is repaired."[25] The circulation of blood in the microcosm of the human body — the subject of Hutton's medical dissertation in Leiden many years earlier[26] — fitted perfectly into this metaphor of the organism, as an analogy no less appropriate than that of a steam engine. Likewise, his meteorology, and in particular his theory of rain, was directed toward elucidating what was well recognized as another process of circulation — in modern terminology, the hydrological cycle: "All the surface of this earth is formed according to a regular system of heights and hollows, hills and valleys, rivulets and rivers, and these rivers return the waters of the atmosphere into the general mass, in like manner as the blood, returning to

23 Ibid., 210.

24 Ibid., 216.

25 James Hutton, *Theory of the Earth, with Proofs and Illustrations,* Vol. 2 (Edinburgh: Cadell, Junior, Davies, and Creech, 1795), 562. See also Hutton, "Theory of the Earth," 214–16, and Hutton, *Theory of the Earth,* 1:13–17.

26 Arthur Donovan and Joseph Prentiss, "James Hutton's Medical Dissertation," *Transactions of the American Philosophical Society* 70, no. 6 (1980): 3–57.

the heart, is conducted in the veins."[27] The peculiar construction of the earth, then, was that it contained parts that were interdependent, serving as both means and ends of one another, and together constituting a dynamic relationship that allowed the earth as a whole to maintain itself. Like Immanuel Kant's conception of what constitutes an organism, the earth, according to Hutton, consisted of elements whose interaction amounted to the same sense of an internally directed self-organization. Rather than posing a threat to the stability of the earth, perpetual alteration was precisely what allowed it to renovate itself by offsetting erosive forces.

§

Notably, Hutton was writing at a period when geology was just coming into its own as a science, and although he lacked expertise in the field, his contribution lay rather in providing this emerging scientific discipline with a dynamic scheme by connecting and synthesizing the geological with the chemical, the biological, and even the technological into an organic whole. For the sake of such an overarching "great purpose" of the earth, however, any notion of a solid and static structure of the planet had to be sacrificed. Because the essential feature of the earth, Hutton observed, was not to be found in its elements, but rather in their interaction: the fertility of the soil depends upon the loose and incoherent state of its materials, and these materials, in turn, are exposed to the effects of the water and wind, a process of erosion possible only insofar as there is a continual restoration of the earth's crust, and so on. In a passage that implicitly invoked the principle of self-organization, Hutton wrote about the earth in terms that call to mind the idea of homeostasis:

> To acquire a general or comprehensive view of this mechanism of the globe, by which it is adapted to the purpose of being a habitable world, it is necessary to distinguish three

27 Hutton, *Theory of the Earth*, 2:533.

> different bodies which compose the whole. There are, a solid body of earth, an aqueous body of sea, and an elastic fluid of air. It is *the shape and disposition* of these three bodies that form this globe into a habitable world; and it is *the manner* in which these constituent bodies are adjusted to each other, and *the laws of action by which they are maintained in their proper qualities and respective departments,* that form the Theory of the machine which we are now to examine.[28]

From this point of view, the constitution of the earth must be understood as a system of semiautonomously shaped units—within which, as we can see in this passage, Hutton included at least the lithosphere, the hydrosphere, and the atmosphere—resulting from their inner determination under the influence of environmental conditions. It is only by investigating the dynamic interaction between the parts that we are afforded the possibility to understand the whole, and such an inquiry requires a conception of causality beyond that of the linear cause and effect of mechanism. As argued by Hutton, it is in the analysis of the "shape and disposition" of the earth's constituent bodies, and "the manner in which these constituent bodies are adjusted to each other," such that their "proper qualities" are but temporally maintained states, that we come to an understanding of what constitutes the earth as a systematic whole.

The Disappearance of History in Deep Time

Hypothesizing an indefinite number of strata perpetually being produced by the planetary heat engine, Huttonian geology would first and foremost have a significant influence on the cultural imagination of historical time. Referring to the existential effect of their attempts to decipher the stratified layers of rock at the outcrops of Siccar Point, Hutton's colleague and travel companion John Playfair colorfully described how his

28 Hutton, "Theory of the Earth," 211 (my emphasis).

"mind seemed to grow giddy by looking so far back into the abyss of time."[29] Although the phrase "deep time" has its origin in nineteenth-century literature,[30] the notion of an abyssal past beneath our feet first emerged with Hutton's geotheory in the late eighteenth century. For if it was from Nicolas Steno's 1669 postulation of the Stratigraphic Law of Superposition that depth first acquired a temporal meaning, then it was upon the basis of this idea that Hutton's theory of infinitely repeating cycles of deposition and erosion—powered by a self-propelling heat engine at its core—freed the discipline of natural history from foundationalism by radicalizing the depth of the telluric netherworld into that of an unfathomable void. Under the layers of granite are further strata of slate, Hutton noted, and were we theoretically to proceed all the way down to the most foundational strata in their hardened state, he speculated, we would find that not even these constitute in any sense an origin or a beginning to the natural history of the earth. Hutton, in the earliest draft of his geotheory, said, "With respect to human observation, this world has neither a beginning nor an end."[31] Likewise, he concluded the full version of his paper by claiming only that this is a limit to what humans can find. It is not entirely clear whether these Kantian insinuations were derived from the careful wording on Hutton's part in order to avoid accusations of impiety. But given his concern, in *An Investigation of the Principles of Knowledge* (1794), with establishing a sound basis for rational understanding, it is unsurprising that he phrased his eternalism with regard to the limitations of human knowledge. For one thing, Hutton denied, out of principle, the validity of Jean-André Deluc's claim to have measured the age of the con-

29 Playfair, "Biographical Account of the Late Dr. James Hutton," 73.

30 Thomas Carlyle, "Boswell's Life of Johnson," in *Macaulay's and Carlyle's Essays on Samuel Johnson,* ed. William Strunk, Jr. (New York: Henry Holt, 1895), 139.

31 James Hutton, *Abstract of a Dissertation Read in the Royal Society of Edinburgh, Upon the Seventh of March, and Fourth of April, MDCCLXXXV, Concerning the System of the Earth, Its Duration, and Stability* (Edinburgh: Royal Society of Edinburgh, 1785), 28.

tinents, since, as he wrote, "it is in vain to attempt to measure a quantity which escapes our notice and which [human] history cannot ascertain."[32] Yet, such a rejection of geogony on epistemological grounds is clearly insufficient for understanding the ontological implications of deep time. Although Hutton held that it was impossible to empirically demonstrate that the earth had a beginning or that it would have an end,[33] to not interpret this statement ontologically would be to neglect the whole purpose of the "wisely designed machine" that he had taken such pains to reveal.

Metaphysically—or, more precisely, ontotheologically—Hutton could not bring himself to believe that the earth, since it was constantly regenerating, existed for no other purpose than to maintain itself eternally. In fact, he could not conceive that the earth, precisely because of the circularity of its operation, could ever truly have a beginning or an end. "The natural course of time, which to us seems infinite," he wrote, "cannot be bounded by any operation that may have an end, the progress of things upon this globe, that is, the course of nature, cannot be limited by time."[34] However vast its putative timescale, then, nothing could have been more profoundly ahistorical. The abyss of deep time opened up by Hutton's geotheory was the gaping chasm of a temporality so vast as to call into question the very term "history." Rather than moving backward to a remote past even tentatively attainable as an origin, or forward to some definite goal waiting to be actualized, this trajectory operated through a dynamically static self-organization intrinsic to the very production of the earth by and for itself.[35] Indeed, Hutton showed no interest in plotting the particularities of the history of the earth. The primary schist and the secondary strata were important only as instances of passing phases in an ahistorical

32 Hutton, "Theory of the Earth," 298.
33 Hutton, *Theory of the Earth*, 1:94, 1:372.
34 Hutton, "Theory of the Earth," 215.
35 Stephen J. Gould, *Time's Arrow, Time's Cycle: Myth and Metaphor in the Discovery of Geological Time* (Cambridge: Harvard University Press, 1987), 86–91.

regularity.[36] Even the successive worlds that he inferred from the rocks at Jedburgh were of significance to him only as evidence of the reproduction of a steady state. Analogous to the harmonious balance between uplift and erosion, which guaranteed an eternal succession of worlds, Hutton's cyclical conception of the earth system took the Newtonian model of a fundamentally stable solar system, with a perfectly balanced motion in the orbit of the planets, as its inspiration. But whereas Newton's cosmos was infinite in space, Hutton, with reference to the dynamic equilibrium between the earth's creative and destructive forces, rather emphasized the infinity of time. In fact, having just reaffirmed the infinite activity of our planet's self-organization, this is precisely how he prefaced the concluding lines to his *Theory of the Earth*—by drawing the connection, once more, between the eternal return of new worlds upon the face of the earth and the eternal return of the planets to their initial starting point in their revolution around the sun: "We have the satisfaction to find, that in nature there is wisdom, system, and consistency. For having, in the natural history of this earth, seen a succession of worlds, we may from this conclude that there is a system in nature; in like manner as, from seeing revolutions of the planets, it is concluded, that there is a system by which they are intended to continue those revolutions."[37]

To use Hutton's own comparison, his successive worlds were as unspecific as the successive orbits of the planets around the sun: events with temporality but without history. In fact, Hutton's eternalist claim has been almost exclusively interpreted as a product of his deism. For instance, Rudwick holds that Hutton's deism not only runs through all his writings but is fundamental to his entire intellectual project of demonstrating that the world displays systematicity—in the sense of designful purpose—such that any appearance of accident or disorder is deceptive.[38] Hence, to discover the system or order of the natural

36 Rudwick, *Bursting the Limits of Time*, 172.
37 Hutton, "Theory of the Earth," 304.
38 Rudwick, *Bursting the Limits of Time*, 160–61.

world was the underlying goal of all of Hutton's writings in natural history, which means that his geotheory too is unintelligible except in the light of his deistic beliefs.

Indeed, Hutton was so adamant in his critique of Mosaic chronology that he has been retroactively ascribed, by such luminaries as Edward Battersby Bailey, the venerable title of "the father of modern geology,"[39] an attitude that has only been underscored because the currently estimated figure of 4.5 billion years has proven correct Hutton's claim that the age of the earth is far beyond human imagination. In the eighteenth century, however, there were no means by which to measure such vast expanses of deep time. This only became possible when the process of radiometric dating was refined in the early twentieth century, which meant that earth scientists could determine the age of rock with reference to the rate of radioactive decay. Instead, at the time Hutton was writing, the authoritative source on the matter was the seventeenth-century calculations of the Archbishop James Ussher of Armagh, who had inferred from the chronology of the Bible that the earth was created in the year 4004 BCE.[40] Only in 1778, less than a decade before Hutton had an abstract of his geotheory presented to the Royal Society of Edinburgh, had Comte de Buffon stuck out his head to become the first naturalist to use the scientific method to calculate the age of the earth, ingeniously assuming that the planet had started out as a ball of molten rock that subsequently cooled down to its current temperature. Having spent six years measuring the cooling rate of materials in his laboratory,[41] he then calculated that the earth was approximately 75,000 years old.[42] Buffon's premise was clever, if flawed. Despite his conservative

39 Edward B. Bailey, *James Hutton: The Founder of Modern Geology* (Amsterdam: Elsevier, 1967).

40 Rudwick, *Earth's Deep History*, 11–14.

41 Jan A. Zalasiewicz et al., "Introduction: Buffon and the History of the Earth," in G.-L. Leclerc, *The Epochs of Nature*, trans. and eds. Jan A. Zalasiewicz, Ann-Sophie Milon, and Mateusz Zalasiewicz (Chicago: University of Chicago Press, 2018), xxii.

42 Leclerc, *The Epochs of Nature*, 35.

estimation, his efforts nevertheless marked a paradigm shift in geogony through the manner by which he proceeded, turning, as he did, to nature itself rather than to scripture. It is in the same vein as Buffon that Hutton would make his own predictions, with a clear refusal to submit to revealed knowledge. But assuming that Hutton therefore provided a secular challenge to biblical natural history would be to overlook the religious views that essentially informed his understanding of our planet as a self-organizing and self-perpetuating "earth-machine." For Hutton as for Ussher, the earth was rationally organized and thus a testament to God's goodness, an earth whose perfection was often, and circularly so, both inferred from the fundamental goodness of the Creator and taken as a proof of it.

Although the emerging natural sciences required a methodological distinction from theology, the two certainly did not become completely severed from each other overnight. In the early eighteenth century, spurred on by a growing antipathy toward more orthodox forms of Christianity, natural science instead became the new site for revelation, gradually replacing scripture as the basis for truth while retaining the scholastic quest to find unity and truth in the diversity of phenomena, and deism was symptomatic of this growing trend. Ever since the days of Francis Bacon, naturalists on the British Isles had succeeded in linking empirical and experimental scientific practice with Christian faith. In fact, the Christian worldview constituted the unifying intellectual framework for natural science, and scientific discoveries were introduced into the general culture as evidence of God's purposeful design. Before the turn of the nineteenth century, philosophy and science were largely interchangeable. But to a greater extent, philosophy came to be associated with theology and metaphysics, while science was linked to mathematics, experimental research, and especially physics. The fact that the Baconian method eventually succeeded in monopolizing the concept of science marks a reversal of the positions in the hierarchy of knowledge, and this reversal pertained to geology too.[43]

This historical circumstance accounts for the currency of Christian trope of the "Book of Nature" around the turn of the eighteenth century, rendering the stratified registration of different rock types into lines of legible text within which were written the motive forces of the earth's history.[44] As laid down by Playfair, the perpetuity of the earth was ultimately guaranteed and maintained by God's goodness, for the Creator "has not given laws to the universe, which, like the institutions of men, carry in themselves the elements of their own destruction. He has not permitted in His works any symptoms of infancy, or of old age, or any sign by which we may estimate either their future or their past duration."[45] The earth, exhibiting delicate mechanisms of interaction between its many elements, had so clearly been created for the benefit of the creatures living upon it that it would have been unfit for such a sophisticated design to have materialized into anything else than a perpetual motion machine. A product of God's wisdom, the planetary heat engine would assuredly continue to perpetuate the cycles of uplift and erosion endlessly, thereby maintaining the conditions necessary for human flourishing, and at the same time it was this indefiniteness that was interpreted as evidence for divine creation. When Hutton's geotheory was first publicly presented, then, those among the audience who were deists probably realized that it accorded nicely with their own beliefs. Throughout the eighteenth century, largely in France, deists had steadily attempted to undermine a literal adherence to the book of Genesis by stressing the inadequacy of the Noachian flood to explain fossils, the discrepancy between the supposed days of creation and geological periods, and the sufficiency of natural causes

43 Rudwick, *Bursting the Limits of Time,* 55–58.

44 Noah Heringman, *Romantic Rocks, Aesthetic Geology* (Ithaca: Cornell University Press, 2004), 188.

45 John Playfair, *Illustrations of the Huttonian Theory of the Earth* (Cambridge: Cambridge University Press, 2011), 119.

as opposed to special creations or other miracles.[46] Like many deists, Hutton did not believe that life had been created serially, but he did not believe in extinction either. However, orthodox Christians would have taken the emphasis on deep time, both in Hutton's *Theory of the Earth* and in his earlier *Abstract of a Dissertation Read in the Royal Society of Edinburgh*, for what it was: an implicit rejection of the creationist narrative and an assertion of total reliance upon what Hume called "natural religion." In fact, eternalism was commonplace among Enlightenment intellectuals, stemming from the prevailing view of nature as a well-oiled machine in which every animal and plant had a divinely allotted place, constituting a harmonious balance in nature. This was manifested in the prevalent belief that there are no unjustifiable absences in existence, or, no things that could be but simply never are without any further justification, since arguing that nature has no unjustifiable gaps is the same as to argue that nature is as justifiable as it can possibly be.[47] Each animal and plant had its God-given purpose, and fed on each other accordingly, in just the right numbers to keep the balance forever stable.

Such steady-state models based on the dynamic interaction of opposed powers were commonplace in Enlightenment thinking. Similar to Hutton's depiction of the circularity of the earth's geological economy, with just the right balance between the creation and destruction of landmass, his good friend Adam Smith's economic theory of societal self-organization constituted yet another contemporary example of an application of the steady-state model—in Smith's case, the intrinsic telos of an invisible hand to govern the interaction between producers and consumers in a decentralized market. Contrary to how it had been conceptualized in the history of economics before him, Smith saw society as fundamentally dynamic, with internal checks and balances emerging from the clash of forces. In

46 Rudwick, *Bursting the Limits of Time*, 55–58.

47 Arthur O. Lovejoy, *The Great Chain of Being: A Study of the History of an Idea* (Cambridge: Harvard University Press, 1936).

addition to the inspiration that Hutton took from the regularity of the laws of motion that Newton had formulated in his *Principia* (1687), here was thus another analogy between systems: as the forces of creation and destruction of landmasses cancel each other out over the course of deep time by naturally settling for an equilibrium in the geological economy, so societies too fluctuate dynamically around an average with respect to total wealth. As a result, this aspect of Hutton's theory came as no surprise to his readers, for natural theology had long emphasized the significance of systems that maintained themselves within a certain range of constancy.

§

This illustrates the extent to which Enlightenment ideals had been cemented in the minds of contemporary geotheorists such as Hutton, which put them on a collision course with the biblical account of natural history. First and foremost, it illustrates the emergence of the modern tension between, on the one hand, the rationalist confidence in the faculty of reason to render a unifying perspective on nature comprehensible to humans through a combination of theory and experiment, and, on the other, the lingering reverence for the sublimity of nature that, it was believed, could not be fully captured by the kind of methodological restriction that mechanically dissected and treated its manifestations as nothing more than dead objects. Driven by his desire to uncover the earth's hidden operations, so as to ascertain the rational system inherent to nature, Hutton's figuration of the geologist as something of a physico-theological pioneer, with his ability to decipher signs of the divine inscribed in geological strata, is part of the emergence of a general discursive formation during the modern period—an equal part religious as it was scientific, and an equal part awestruck by the immense complexity of nature as by the technological power and sophistication that had made such an intimate understanding possible in the first place. Retelling his 1786 tour of the southern Uplands of his native Scotland, Hutton depicted himself and his companion John Clerk as resolute trailblazers making their way on "a road which perhaps was never passed in a chaise before."[48] Impatiently, the two pressed on, eager to discover whether the outcrops near Solway Firth promised to deliver the kind of stratified rock from which may be derived the purposeful order of the earth's constitution:

> Breaking through the bushes and briars, and climbing up the rocky bank, if we did not see the apposition of the granite to the side of the erected strata so much as we wished, we saw something that was much more satisfactory, and to the purpose of our expedition. This was the granite superinduced upon the ends of those broken strata or erected schisti. We now understood the meaning of the impending granite which appeared in the hill above this place; and now we were satisfied that the schistus was not only contiguous with the mass of the granite laterally, but was also in the most perfect conjunction with this solid rock which had been superinduced upon the broken and irregular ends of the strata.[49]

Indeed, going through the natural archive of worlds long past, far preceding the record of human history, Hutton appeared to be standing on the threshold of deep time, about to penetrate the outer layer of appearances so as to set out on an intellectual journey to the core of its Plutonic forces — the innermost kernel of the earth itself. Although nature rather than the Bible was the medium, it was nevertheless so that, in both cases, the vestiges of design were intrinsically there to be read. In fact, the rhetoric of the age spoke of *lapides literati* and "graphic granite," in which the history of the ages was recorded.[50] Here we find geology and

---

48 James Hutton, *Theory of the Earth, with Proofs and Illustrations*, Vol. 3 (Edinburgh: Cadell, Junior, Davies, and Creech, 1795), 54.

49 Ibid., 57–58.

50 Theodore Ziolkowski, *German Romanticism and Its Institutions* (Princeton: Princeton University Press, 1990), 33, and Frank D. Adams, *The Birth and Development of the Geological Sciences* (Baltimore: The Williams and Wilkins Company, 1938), 250–76.

philology juxtaposed in a manner that seems far more outlandish to us today than it did to the naturalists working prior to the disciplinary sundering of the gulf between the natural and the human sciences. For in the eighteenth century, the geological and the philological enterprises could still be thought of as analogical: in both cases, a hermeneutic exercise in uncovering the hidden grounds beneath surface meaning. Just like the philologist traces a language given to us in the present back to its origins, so the geologist strips away the sediment to uncover what is primordial in nature. Comparing the hermeneutic faculty required to study dead languages with the method needed for studying the inorganic strata of nature, the task was one of recognizing "the living spirit" in a seemingly dead product, that is, to discover the abyss of productivity underlying it.[51]

The literary theorist Noah Heringman has coined the term "aesthetic geology" to capture this interplay between geological excursion and aesthetic experience during the Romantic period, pointing out that, prior to the formation of geology and literature as mutually exclusive disciplines in the nineteenth century, both were equally considered to belong to the general domain of men of letters.[52] Before solidifying into the guarded province of college professors and scientists, geology was as much of a fascination for men of letters as it was for natural historians, with its subject figuring in the popular press and in periodicals. But more importantly, geology became *the* hermeneutic paradigm par excellence, providing rhetorical symbols for philosophers, theologians, artists, and politicians alike. Thus geologists happily borrowed their metaphors from historians, archivists, and archeologists in declaring their findings to be ecofacts taken straight out of the Book of Nature, but Romantic artists, writers, and poets in turn often ventured into natural history and saw the findings of the famous geologists of their time as an opportunity to speculate on the ontological consequences of

51 Friedrich W.J. von Schelling, *On University Studies*, trans. E.S. Morgan (Athens: Ohio University Press, 1966), 39–40.

52 Heringman, *Romantic Rocks, Aesthetic Geology*, 155.

deep time.[53] Indeed, "the metaphorical structure of Romanic poetry," Northrop Frye remarks, "[tended] to move inside and downward instead of outside and upward."[54] Whereas eighteenth-century poets in Britain and Germany sought out steep heights for their view of the sublime, their Romantic successors opted to follow in the footsteps of the miner rather than the mountaineer, applying their aesthetic sensitivities to the joint effort of probing the speculative richness that lay hidden in the depths of the earth. "German literature of the Romantic age," Theodore Ziolkowski confirms, "is crawling with so many miners [...] that the unwary reader might well believe he had blundered into a surrealistic library in which the history of literature is interspersed at random with the history of technology."[55] Similarly, the cultural repercussions of Huttonian geology really came into their own with the British Romantics in the beginning of the nineteenth century, when they enrolled the affective register of deep time as one of terror and wonder, fashioned to fit an ambivalent vision of the sublime that transcended and yet somehow also affirmed the human. It is no surprise, in other words, that in Percy Bysshe Shelley's great lyrical drama *Prometheus Unbound* (1820), the erupting volcano, so distinctive of Plutonism, is the controlling image.

But as suggested by the very title of Shelley's drama, this is not to argue that the aestheticization of the geophysical domain of nature worked to restrain the technological march of a brewing industrialism. On the contrary, there was a strong economic context to the development of geology as a scientific discipline, linked to coal mining in Britain and the mining of precious metals in Germany, and at the time steeped in a Romantic discourse of the heroic endeavor of exposing, literally, the innermost secrets of the earth. For centuries, the demand for metals had

53 Cécile Roudeau, "The Buried Scales of Deep Time: Beneath the Nation, beyond the Human ... and Back?," *Transatlantica* 1 (2015): 5.

54 Northrop Frye, "The Drunken Boat: The Revolutionary Element in Romanticism," in *Romanticism Reconsidered,* ed. Northrop Frye (New York: Columbia University Press, 1963), 16.

55 Ziolkowski, *German Romanticism and Its Institutions,* 18.

been restricted principally to agricultural and military needs, but as new industrial uses were found for metals — not the least in the modern machine — the demand for ores increased dramatically. At the same time, the coal industry began to expand rapidly because it was discovered that coke rather than charcoal could be used in the smelting and manufacture of iron.[56] "More closely than any other industry," Lewis Mumford argued in *Technics and Civilization* (1934), "mining was bound up with the first development of modern capitalism."[57] Recounting how the Central European mining boom of the fifteenth century led to a widespread demand for the kind of expensive equipment required to access ever deeper ores, the French historian Fernand Braudel recounted how financiers could relatively quickly establish their control over the business. As mine workers found themselves increasingly dependent on investment, "capitalism," Braudel observed, "entered a new and decisive stage[. ...] Indeed," he even notes, "this was when the word *Arbeiter*, worker, first appeared."[58] To be sure, the rise of extractive capitalism had a huge influence on the cultural imagination of the early industrializers of the West, coming to the fore in the Romantic movement. In Germany, for instance, there was an established state mining bureaucracy, such as the international school for mining engineers at Bergakademie Freiberg, where one of the foremost of the early Romantic poets, Novalis, studied under Werner. In fact, it would have been difficult to assemble a group of intellectuals in any of the centers of German Romanticism around the turn of the eighteenth century without including at least one or two guests who were, in one way or another, involved in either

56 Most importantly, the mine is deeply implicated with the emergence of the mode of production that the human ecologist Andreas Malm takes to task for triggering global environmental change, what he calls "fossil capital"; see Malm, *Fossil Capital: The Rise of Steam Power and the Roots of Global Warming* (London: Verso, 2016).

57 Lewis Mumford, *Technics and Civilization* (London: Routledge & Kegan Paul, 1955), 74.

58 Ferdinand Braudel, *The Structures of Everyday Life*, vol. 1: *Civilization & Capitalism, 15th–18th Century* (New York: Harper Collins, 1982), 321–22.

the study or extraction of minerals. For Goethe, too, who for some time during his career as a statesman was put in charge of silver mining in the duchy of Saxe-Weimar, developed a lifelong interest in geology and established friendly contacts with Werner and his colleagues at the Bergakademie. Similarly, before setting out on his scientific exploration to South America, Alexander von Humboldt spent several years as an *Oberbergmeister* in the Prussian Department of Mines.[59] Hence, as an expression of the *Zeitgeist*, the confrontation between nature and artifice played out in the economy, the sciences, and the arts simultaneously, often with an implicit overlap.

Crucial to note, here, is the circulation of ideas back and forth across the categories of the natural and the artificial. Premised upon the ontologically flat concept of a system, there was an effort to secure an underlying harmony between divine guidance, the system of nature, and human industry. Far from particular to the work of Hutton, the aim to supply the sciences of the human with the methods of natural philosophy was common. The political economist Alec Macfie famously identified a similar effort in the work of Hutton's fellow Scotsman and intellectual interlocutor Adam Smith to integrate the theological, jurisprudential, and ethical with the economical through the concept of "the invisible hand," so as to bind them all into one comprehensive system of thought.[60] As is well known, the invisible hand of the benevolent Christian deity appears both in *The Theory of Moral Sentiments* (1759) and in *An Inquiry into the Nature and Causes of the Wealth of Nations* (1776) as a kind of conservative force that gravitates society toward natural order through the interactions between self-interested individuals. But as Macfie notes, the concept also surfaces in Smith's "History of Astronomy" (1795), this time in the form of "the invisible hand of Jupiter,"[61] but then described as a capricious and ener-

59 Ziolkowski, *German Romanticism and Its Institutions*, 19–22.
60 Alec Macfie, "The Invisible Hand of Jupiter," *Journal of the History of Ideas* 32, no. 4 (1971): 595–99.
61 Adam Smith, *Essays on Philosophical Subjects*, eds. William P.D. Wightman, John C. Bryce, and Ian S. Ross (Oxford: Oxford University Press,

gizing force that breaks loose the status quo. However, Macfie is keen to argue that there is no inconsistency in Smith's usages of the concept, for taken together they suggest the same steady-state dynamics of creation and destruction, order and disorder, that we find in Hutton's geotheory, with only a difference on emphasis. Whether natural or human history, both Hutton and Smith sought to uncover the hidden laws governing the empirical world from behind the scenes. Put differently, the same self-organizing principle that structures nature holds true for society, such that what fundamentally governs the geophysical production of the earth's natural economy is found to be identical to the mechanisms behind the industrial system of production and political economy.

Thus, in spite of its radical historical pretenses, Huttonian deep time tended rather toward the universalizing and standardizing logic of the industrial factory, masking the historical particularities of the capitalist mode of production by construing it as an expression of underlying natural law.[62] It was according to this logic, as Simon Schaffer has argued, that the idea of a "natural system," in the wake of the Enlightenment, was rendered ontologically equivalent with that of a "factory system."[63] It is no wonder then that Hutton — one of Smith's literary executors — posthumously oversaw the publication of his essays in the history of science, wherein the concept of system was put to work in such a manner that Smith could find plentiful machines in nature, and, conversely, the mere "continuation" or "perfection" of nature in the industrial machine: "Systems in many respects resemble machines. A machine is a little system, created to perform, as well as to connect together, in reality, those different movements and effects which the artist has occasion for. A system is an imaginary machine invented to con-

1980), 49.

62 Jussi Parikka, *A Geology of Media* (Minneapolis: University of Minnesota Press, 2015), 41.

63 Simon Schaffer, "Babbage's Intelligence: Calculating Engines and the Factory System," *Critical Inquiry* 21, no. 1 (1994): 203–27.

nect together in the fancy those different movements and effects which are already in reality performed."[64]

In a manner identical to that of the geotheorist, Smith reckoned that only those observers concerned with the deeper meaning of the phenomena observed — as opposed to "those of liberal fortunes, whose attention is not much occupied either with business or with pleasure"[65] — could construct systems that made sense of the conduct of everyday labor. Similarly, Hutton stipulated that even though science was undoubtedly useful, only philosophy, in its pursuit of a system of knowledge to integrate the various branches of inquiry into an organic totality, could really satisfy rational humankind.[66] In short, as for the geotheorist, so also for the social theorist. In both cases, what might seem random or miraculous to the uneducated or superstitious observer could nevertheless be accounted for by the enlightened scholar as elemental parts of a rational and providentially planned system, only graspable by those who perceive the parts in relation to the whole.

But even though we are faced with a conceptual exchange across the categories of nature and artifice, there is certainly no acknowledgment, in this lineage of reasoning, of the genuine artificiality of nature as a concept. On the contrary, such is the purview of the historicist. Consequently, it is instructive to note that Darwin, who was an avid reader of Lyell, owed a considerable intellectual debt to the political-economical literature of his time, and not the least to the writings of Smith. In the *Wealth of Nations,* for instance, Smith had imagined a tribe of hunters. One member, he suggested, makes good weapons but is an indifferent hunter, and thus finds that he does better by making weapons and trading them for game than he does by combining this activity with hunting. If other members of the tribe simi-

64 Smith, *Essays on Philosophical Subjects,* 66.

65 Ibid., 50.

66 James Hutton, *A Dissertation upon the Philosophy of Light, Heat, and Fire* (Edinburgh: Cadell, Junior & Davies, 1794), v. See also Roy Porter, "Gentlemen and Geology: The Emergence of a Scientific Career, 1660–1920," *The Historical Journal* 21, no. 4 (1978): 809–36.

larly develop their own strengths, Smith proposed, and even if everyone seeks only their own interest and cares nothing for the general good, their selfishness would still lead to a division of labor that results in increased production.[67] Here, each is "led by an invisible hand to promote an end which was no part of his intention,"[68] namely, the maximization of the productive capacity of the economy as a whole. According to Smith, this phenomenon reaches its apotheosis in industrial capitalism, and in particular in the factory system, to which he attributes an astounding increase in productivity precisely because it seeks to exploit the division of labor as thoroughly as possible.[69] An economy therefore produces the most wealth when individuals compete to achieve their private interests. Without a central planner, the invisible hand guarantees the best economic interests of the community as a whole. Compare this depiction of society to Darwin's theory of evolution. In nature's economy, organisms compete to survive and reproduce, a struggle for existence that is absolutely central to the evolutionary process. In the struggle for existence, as in economic competition, different organisms are better at exploiting different resources, giving rise to a "natural division of labor." Again, as if guided by an invisible hand, "so in the general economy of any land, the more widely and perfectly the animals and plants are diversified for different habits of life," Darwin declared, "so will a greater number of individuals be capable of there supporting themselves."[70] The echoes of Smith are unmistakable.

As has already been observed by a myriad of historians, there is thus a certain irony in Herbert Spencer's enrollment of natu-

67 Adam Smith, *An Inquiry into the Nature and Causes of the Wealth of Nations,* vol. 1, eds. Roy H. Campbell, Andrew S. Skinner, and William B. Todd (Oxford: Oxford University Press, 1976), 28–30.

68 Adam Smith, *An Inquiry into the Nature and Causes of the Wealth of Nations,* vol. 4, eds. Roy H. Campbell, Andrew S. Skinner, and William B. Todd (Oxford: Oxford University Press, 1976), 456.

69 Smith, *Wealth of Nations,* 1:14–24.

70 Charles Darwin, *On the Origin of Species by Means of Natural Selection, or the Preservation of Favoured Races in the Struggle for Life* (London: John Murray, 1859), 115–16.

ral selection into his social Darwinist dictum, "the survival of the fittest," as an ethico-political precept to sanction a social structure built around cutthroat competition à la laissez-faire economics, eugenics, and pseudoscientific forms of racism by means of an appeal to nature, because Darwin himself had already drawn upon Smith's economic theory and only secondarily rendered it a biological fact. In the case of the concept of system, then, the circulation of ideas made possible by its flat ontology was of a one-directional kind, namely, the naturalizing kind. The social Darwinists were certainly guilty of violating Hume's law in their eagerness to derive an "ought" from an "is," but there is something more to be said about the inverse movement from a prescriptive to a descriptive statement — certain natural historical and antihistoricist tendencies observable in the works of Darwin, Smith, and Hutton alike. Karl Marx noted already in 1842 in a letter to Friedrich Engels, commenting on the connection between Darwin's theory of evolution and the bourgeois economics of Smith: "It is remarkable how Darwin rediscovers, among the beasts and plants, the society of England with its division of labor, competition, opening up of new markets, 'inventions' and Malthusian 'struggle for existence.'"[71] Extending Marx's observation into a critique of ideology, Engels would later argue that Darwin's theory amounted to, among other things, and however inadvertently, a naturalization of customary economic arrangements:

> All that the Darwinian theory of the struggle for existence boils down to is an extrapolation from society to animate nature of Hobbes' theory of the *bellum omnium contra omnes* and of the bourgeois-economic theory of competition together with the Malthusian theory of population. Having accomplished this feat [...] these people proceed to re-

71 Karl Marx, "Letter to Friedrich Engels, 18 June 1862," in *The Collected Works of Karl Marx and Friedrich Engels,* vol. 41: *Letters 1860–1864,* eds. Jack Cohen et al., trans. Peter Ross and Betty Ross (London: Lawrence and Wishart, 2010), 381.

> extrapolate the same theories from organic nature to history, and then claim to have proved their validity as eternal laws of human society.[72]

Whereas the German historicist tradition, from Werner to Marx, came to interpret the past as the key to the present, the British empiricist tradition, coming out of Hume's critique of the fallacy of inductive inference, on the contrary saw the present as the key to the past[73] — in Smith's case, basing his economic history upon contemporary human behavior and market development, and in the case of Hutton's geotheory reducing earth history to what was directly observable in the actual landscape.

It is in this sense of its cultural implications that the birth of geology, as a modern scientific discipline, not only provided the emergent mode of industrial production with access to much needed raw materials, but also exercised an extraordinary influence upon the entire literary and philosophical imagination as it pertained to the cultural self-consciousness of humanity's relationship to nature during this revolutionary period of technological transformation.[74] We thus need to be able to understand this ontological equivalence also as the political-economic function of the emerging intellectual genealogy of Huttonian geology. Indeed, this is where geological concerns reveal an entirely other side to conceptual deep time — one that puts the earth at the center of a cultural reevaluation of the human subject and its status in the natural world. As noted by the philosopher and historian of architecture Amy Kulper, the ontological dimension to the intellectual history of late eight-

72 Friedrich Engels, "Letter to Pyotr Lavrov, 12–17 November 1875," in *The Collected Works of Karl Marx and Friedrich Engels*, vol. 45, *Letters 1874–1879*, eds. Jack Cohen et al., trans. Peter Ross et al. (London: Lawrence & Wishart, 2010), 107–8.

73 A.M. Celâl Şengör, *Is the Present the Key to the Past or Is the Past the Key to the Present? James Hutton and Adam Smith versus Abraham Gottlob Werner and Karl Marx in Interpreting History*, The Geological Society of America Special Paper, Vol. 355 (Boulder: The Geological Society of America, 2001).

74 Ziolkowski, *German Romanticism and Its Institutions*, 18–62.

eenth- and early nineteenth-century geology was in fact crucial for inventing new ways to make sense of the technological extraction of value from nature on an industrial scale. For "as a result of eighteenth-century archeological and antiquarian activities, the earth acquired a new perceptual depth, facilitating the conceptualization of the natural as immanent history, and of the earth's materials as resources that could be extracted just like archeological artifacts."[75] Such a poietic domestication of nature took geology as its model and proposed to liberate nature from instrumentalism by reinscribing spirit into natural history, in effect rejuvenating humankind's relationship to the earth by the means of a discursive formation based upon the notion of artistic production. Precisely insofar as they participate in the perfection of spirit in nature, humans would come to possess an authentic relationship to their terrestrial abode by taking up a poietic use of its materials. "This aesthetic response to the materiality of rocks and landforms," Heringman argues, "is inseparable from the emerging economic category of natural resources," since the conflation of the "rock record" into both literary and scientific metaphor provided the aesthetic terminology necessary to imagine the technological enrollment of nature into human industry without reducing it to the instrumentalism of economic utility, thereby preserving "the paradox of a landscape both profitable and 'romantic.'"[76] In this manner, human artifice too could be depicted as part of a comprehensive archive of natural history.

A Natural History of Artifice

Social turmoil and intellectual paradigm shifts are seldom isolated from rapid technological change. Pinpointed as the onset of the Anthropocene, the late eighteenth century marks

75 Amy C. Kulper, "Architecture's Lapidarium: On the Lives of Geological Specimens," in *Architecture in the Anthropocene: Encounters among Design, Deep Time, Science and Philosophy*, ed. Etienne Turpin (Ann Arbor: Open Humanities Press, 2013), 100.

76 Heringman, *Romantic Rocks, Aesthetic Geology*, 161.

the birth of the industrial age.[77] At the same time as Hutton's geotheory was being laid out on paper, Western societies were striding forward along a trajectory of confidence and improvement. Presaging the idea of humanity as a geological force, Buffon declared already in 1778, albeit on a positive note, that "the entire face of the Earth today carries the imprint of the power of man, which, though subordinate to that of Nature, often created more than did she, or at least marvelously assisted, so it is with the help of our hands that she developed in all her extent, and that she arrived by degrees to the point of perfection and magnificence that we see today."[78]

Contemporaneous with Hutton and his peers, the upheavals in North America and France were similarly spurred by attempts to overturn religious doctrines of divine right as a source of legitimate rule. However, we would be wrong to simply assume that his geotheory was an inherently revolutionary one, at least in the critical emancipative sense inspired by the political ideas emerging out of the Enlightenment. For as would become increasingly clear in Lyell's classic account of Huttonian geology, it rather performed a double movement across the division between human and earth history by reducing the potentiality of the latter to the actuality of the former, all the while privileging the natural structure of the latter over the cultural contingency of the former. In Lyell's hands, the historical significance of deep time was thus further sedimented as of an entirely different kind from that which underlay the emergence of the program of historicism in the nineteenth century, privileging—the latter did—the hermeneutic world of human affairs and, in effect, the contingency of meaning-making practices.

Indeed, in its uniformitarian expression, deep time served to methodologically anchor the possibility of synchronically infer-

77 Paul J. Crutzen and Eugene F. Stoermer, "The 'Anthropocene,'" *IGBP Global Change Newsletter* 41 (2000): 17–18, and Will Steffen, Paul J. Crutzen, and John R. McNeill, "The Anthropocene: Are Humans Now Overwhelming the Great Forces of Nature?," *Ambio* 36, no. 8 (2007): 616–17.

78 Leclerc, *The Epochs of Nature*, 124.

ring from the actual the universal laws that govern the earth prior to any appearance of natural and artificial products in time—or, as methodologically summed up by Hutton himself, "we must read the transactions of times past in the present state of natural bodies."[79] From the point of view of Huttonian geology, then, natural structure could not be considered but interminably identical with rational justification, and as such there could be no real stakes, and thus no true disaster in nature. In Whewell's words, both Hutton and his champion, Lyell, were interested in combating hypothetical reconstructions of great changes "of a kind and intensity quite different from the common course of events, and which may therefore be properly called *catastrophes*,"[80] that is, the kind of abrupt extinction events and violently destructive scenarios of a global dimension that characterized the catastrophism of Georges Cuvier. For the uniformitarians, disaster was a mere epiphenomenon floating over the Plutonic guarantee of system-wide upcycling and the eventual return to a steady state. Even something as seemingly unaccountable as species extinction could be seen as having a reason in the conviction that such apparent deviations from the rational was ultimately qualified and conditioned—and thus justified—by the self-organization of the geosomatic whole.[81] Even when geologists in the Huttonian vein could no longer deny the evidence that organisms had previously gone extinct, or that the technological power of human industry, "differing in kind and energy from any before in operation,"[82] appeared to put the assumed uniformity of nature into question, the same set of ideas nevertheless remained irresistible to such illustrious figures as Lyell, who confidently declared, as late as in the 1830s, that "should there appear reason to believe that certain agents have, at particular periods of past time, been more potent

79 Hutton, *Theory of the Earth*, 1:373.

80 William Whewell, *History of the Inductive Sciences from the Earliest to the Present Times*, vol. 3 (London, John W. Parker, 1837), 606.

81 Thomas Moynihan, *X-Risk: How Humanity Discovered Its Own Extinction* (Falmouth: Urbanomic, 2020), 163–73.

82 Lyell, *Principles of Geology*, 1:243.

instruments of change over the entire surface of the earth than they now are, it is still more consistent with analogy to presume, that after an interval of quiescence they will recover their pristine vigour, than to imagine that they are worn out."[83] Phenomena such as anthropogenic environmental change, in other words, were mere local aberrations from the fundamental vitality of the terrestrial body, understood as a whole and on a geological timescale.

So, although the modern discipline of geology can be said to have begun proper with the late eighteenth-century demythification that sundered "world" from "earth" — in the wake of Hutton's declaration of "having, in the natural history of this earth, seen a *succession* of worlds"[84] — this is not the same as to argue that the opening up of surface appearances to the deeper elemental forces of its fiery core necessarily rid geology of the desire to uncover, beneath, an underlying harmony. For Huttonian geology may well sunder world from earth — surface from depth, appearance from reality — all the while still retaining the primacy of appearances, precisely because infinite depth, at the end of the day, equals complete depthlessness. It was in accordance with such an infinite depth that it would later become appropriate for some of the Romantics to conceive of our planet's geological destiny as embodied in human intention, assuring that even the tiniest cracks in what we took as a solid foundation upon which human history can safely progress were circumscribed as but natural inevitabilities in the much more fundamental and traumatic experience of deep time.[85] In the introduction to the *Philosophy of Nature,* Georg W. F. Hegel wrote, attributing the phrase to his fellow traveler Friedrich W. J. von Schelling, that

83 Ibid., 251–52.

84 Hutton, "Theory of the Earth," 304 (my emphasis).

85 Comparing history to a theatrical drama guided by "an unknown hand," Schelling metaphysically affirmed a general "spirit who speaks in everyone," composing this geocosmic performance on the terrestrial stage as "a progressive [...] revelation of the absolute"; see Friedrich W.J. von Schelling, *System of Transcendental Idealism,* trans. Peter Heath (Charlottesville: University of Virginia Press, 1978), 209–10.

before "the stones cry out and lift themselves up to spirit," nature remains a "petrified intelligence."[86] Having ontologized Huttonian deep time in order to make the point that geogony is "nothing other than a natural history of our mind,"[87] Schelling had called this ontological basis of history "the past-in-itself,"[88] and pointed out that insofar as one tries to include this geophysical source of the subject in the schematism of the mind's catalogue, it retracts infinitely from intentional consciousness by introducing an "indivisible remainder": a surplus unassimilable to cognition, yet, Schelling believed, therefore also productive of it.[89] As opposed to the post-Freudian reception of the unconscious as an exclusively psychological and brain-bound domain, the German Romantic version was that of a radically geophysical phenomenon, taken to permeate the entirety of nature and in effect attributed specifically to abiotic processes.[90]

Perfectly in line with the archival discourse among natural historians of the eighteenth century,[91] the geocosm was depicted by the Romantics as "the unconscious," and the strata as its geophysical memory bank. Though the living spirit speaks in language, nature was imagined as the author of a book that must be translated with the skill of a "natural philologist," that is, a stratigrapher. Drawing from Goethe's poem "The Apotheosis of the Artist" (1789), Schelling noted how "nature is like some very ancient author whose message is written in hieroglyphics

86 Georg W.F. Hegel, *Hegel's Philosophy of Nature,* vol. 1, ed. and trans. Michael J. Petry (London: Allen and Unwin, 1970), 206. See also Alison Stone, *Petrified Intelligence: Nature in Hegel's Philosophy* (Albany: SUNY Press, 2005).

87 Friedrich W. J. von Schelling, *Ideas for a Philosophy of Nature,* trans. and ed. Errol E. Harris and Peter Heath (Cambridge: Cambridge University Press, 1988), 30.

88 Schelling, *System of Transcendental Idealism,* 119.

89 Slavoj Žižek, *The Indivisible Remainder: On Schelling and Related Matters* (London: Verso, 1996), 29.

90 Matt Ffytche, *The Foundation of the Unconscious: Schelling, Freud, and the Birth of the Modern Psyche* (Cambridge: Cambridge University Press, 2012).

91 Heringman, *Romantic Rocks, Aesthetic Geology,* 64.

on colossal pages[. ...] The earth is a book made up of miscellaneous fragments dating from very different ages. Each mineral is a real philological problem."[92] Consequently, the idea of an unconscious memory first emerged as a curious synthesis of the self-identity of absolute idealism with the geological discovery of its radically inorganic and natural historical anteriority. For although the former has its roots in the ancient Greek stipulation of all existents as infinitely contained within the rational structure of ideation's inclusive schema, this collided with the latter's emendation of nature's inorganic antecedence and its postulation of an abyssal geophysical provenance prior to the emergence of the modern subject.[93] Although it is perhaps Ernst Bloch who most famously diagnosed German Romanticism's inorganic unconscious as an immune response against the encroaching threat of death inaugurated by the modern scientific world picture,[94] already Cuvier had in fact acknowledged the survival of this organic prejudice whereby nineteenth-century naturalists could continue, well after the institutionalization of the modern scientific world picture, to harbor the ontological conviction that "all the solid parts of the earth owe their birth to life" such that "each of its parts is alive."[95] In any case, what Bloch noted was that, though habitats and lives were threatened by environmental catastrophe, and although national foundations were cracking, the dissolution brought about by the steam engine upon humanity's environment could, precisely with the help of this organic prejudice, be made to figure in terms of as fundamental an aspect to the regenerating vigor of the earth's unconditioned productivity as the sluggish drift of its tectonic plates. In a geopoetics of disintegration and revolt, destruction

---

92 Schelling, *On University Studies*, 40.

93 Thomas Moynihan, *Spinal Catastrophism: A Secret History* (Falmouth: Urbanomic, 2019), 71–80.

94 Ernst Bloch, *The Principle of Hope*, 3 vols. (Oxford: Blackwell, 1986), 3:1153.

95 Martin J.S. Rudwick, *Georges Cuvier, Fossil Bones, and Geological Catastrophes: New Translations and Interpretations of the Primary Texts* (Chicago: University of Chicago Press, 2008), 201.

could be rendered invigorating, and the cracks themselves part of its form and part of its meaning.

To plant human activity—and grief—in deep time is thus to reinnovate the elegiac by retrieving from its heart the ambivalence of an equal cause for celebration. At the same time as these Romantics acknowledged the immense power of an impending industrial civilization, they did so only by weaving human artifice back into the immanent web of the earth's geophysical forces. Viewed against the background of deep time, the mechanical heat engine did not only appear as a quantitative increase in power and precision in the lineage of "a single, complex pyrotechnic tradition"[96] that spans over 10,000 years and includes everything from ceramics, to metallurgy, to the art of glassmaking. Even more radically, by disclosing human artifacts as mere products of the earth's self-organization, deep time was employed in such a manner as to entirely do away with the ontological distinction between nature and artifice that any history of technology is premised upon.[97] Suddenly, the history of technology is conflated with earth history: metals and chemicals get deterritorialized from the sublimity of geological stratification so that they may instead be reterritorialized in accordance with the "abstract machines"[98] of an unrestrained instrumentalism.

Flattening the Earth

There are a number of implications of Hutton's geotheory that should strike us as remarkable. First, the idea that the earth is a self-organizing body. Second, that he deduced the principle of self-organization—that the earth must possess some fundamental operation of regeneration by which its elements dynami-

96 Theodore A. Wertime, "Pyrotechnology: Man's First Industrial Uses of Fire," *American Scientist* 61, no. 6 (1973): 676.

97 Nigel Clark, "Fiery Arts: Pyrotechnology and the Political Aesthetics of the Anthropocene," *GeoHumanities* 1, no. 2 (2015): 266–84.

98 Gilles Deleuze and Félix Guattari, *A Thousand Plateaus: Capitalism and Schizophrenia*, trans. Brian Massumi (Minneapolis: University of Minnesota, 1987), 63–74; also see the book as a whole.

cally interact so as to generate order out of disorder — largely on ontotheological grounds. Admittedly, Hutton invoked fieldwork to make his case, but these findings were drawn from a surprisingly small number of observations and within a geographically limited area, and the evidence was more often than not selectively chosen to suit his model.[99] Of course, his intuition that Plutonic forces were not one-sidedly destructive found support in the geoscientific community's adoption of the theory of plate tectonics, but that was not until well into the middle of the twentieth century. So, during Hutton's time, no such evidence was yet available, which meant that he instead had to work backward, from the natural theological attitude that became popular among intellectuals during the Enlightenment, to argue that we may, with the tools of reason, "obtain a purpose worthy of the great power that is apparent in the production of it,"[100] that is, to deduce a first principle from which the function of the earth can be explained. This is merely a roundabout way of positing an intrinsic telos to the earth — one that resides naturally rather than supernaturally. Although he did not use the word "God," his conclusions were explicitly motivated by the sense of, as he called it, an "underlying wisdom" in nature. Since the remarkable complexity that we are able to witness on the earth points to such a "delicate contrivance," we may safely conclude that natural products are unlikely to be the outcome of mere blind mechanism.[101] It is in this sense that the crucial difference between Hegel's and Hutton's respective actualisms can be said to lie in the assertion of the latter that "the oldest rocks [are ...] the last of an antecedent series"[102] — an antecedence that Hegel

99 Gould, *Time's Arrow, Time's Cycle*, 60–79; Rachel Laudan, *From Mineralogy to Geology: The Foundations of a Science, 1650–1830* (Chicago: University of Chicago Press, 1987), 128–29; and David R. Oldroyd, *Thinking about the Earth: A History of Ideas in Geology* (Cambridge: Harvard University Press, 1996), 92–96.

100 Hutton, "Theory of the Earth," 209.

101 Ibid., 209–14, 216–17, and James Hutton, *An Investigation of the Principles of Knowledge, and of the Progress of Reason, from Sense to Science and Philosophy*, vol. 1 (Edinburgh: Strahan and Cadell, 1794), 3–13, 17–18.

102 Hutton, "Theory of the Earth," 216.

sought to eliminate. If, as Hutton believed, geology exposes an ungroundedness at the core of the earth, then this is because there is no primal layer or final substrate upon which it ultimately rests. The earth is not an object containing its ground within itself, but more like a never-ending process of grounding — a process that, he argued, is without a definite foundation, an ungrounded abyss.[103] Accordingly, humans cannot recover the origins of the earth since the depth of its crust is contingent in relation to the immanently unfolding process that is its self-sculpting. Such an ungrounding — the opening up of the earth's products to the fundamental turbulence anterior to, or, befittingly enough, beneath them — is like a Plutonian version of the Neptunist deluge: rock flows, just like water, only much more slowly.

The geographer Charles Withers has argued that the idea of humanity's participation in the perfection of the earth's fertility and repair is the crucial link that connects Hutton's early interest in farming and husbandry to his later dabbling in geology. The embodiment of a "gentleman scientist,"[104] Hutton was, like many intellectuals at the time of the cosmopolitan Enlightenment philosophers, something of a polymath. The son of a wealthy Scottish merchant, he supplemented his already ample income primarily by producing chemical compounds. It was while carrying out experiments on the family estates that Hutton had become obsessed with how the geospheres shaped the land on which he worked, and how the progress of new forms of rural management ultimately rested on advancing humanity's understanding of the natural world. But he had no formal training as a geologist, and his work in the field constituted but a small subset of his intellectual endeavors.[105] Insofar as he sought to

---

103 Iain Hamilton Grant, "Mining Conditions: A Response to Harman," in *The Speculative Turn: Continental Materialism and Realism,* eds. Levi Bryant, Nick Srnicek, and Graham Harman (Melbourne: re.press, 2011), 44–45.

104 Porter, "Gentlemen and Geology," 813–15.

105 For Hutton also wrote and published as widely as in medicine, meteorology, agriculture, natural philosophy, and epistemology. In fact, by the time the first two volumes of his *Theory of the Earth; with Proofs and Illustra-*

improve agricultural practice through the scientific method and in effect to transform the eighteenth-century agrarian economy by subjecting it to rational principles — which required grasping the political economy of humankind as part of the geological economy of nature — Hutton's geotheoretical work can be understood as rather conventionally positioned within the intellectual movement of the Scottish Enlightenment.[106] Hence, Bailey has suggested that Hutton's proposal to conceive "of the Earth as of a well-managed agricultural estate with a rotation designed to maintain continuing fertility"[107] indicates not only "the presence and efficacy of design and intelligence in the power that conducts the work"[108] of nature, but, moreover, that humans, through science and technology, can grasp the inherently rational structure of natural systems and thus organize their artificial systems accordingly.

Although he was originally trained as a physician, Hutton had also studied agriculture in East Anglia, France, and Flanders for two years in the middle of the eighteenth century, and then spent an additional thirteen years carrying out experiments and tinkering with agricultural improvements on two estates that he had inherited from his father, the findings of which he summed up in two volumes of an as yet unpublished treatise, titled *Ele-*

---

tions had appeared in 1795, Hutton had already published *Dissertations on Different Subject in Natural Philosophy* (1792), a large quarto volume of essays dealing with meteorology, the nature of fire, and the nature of matter; a smaller but still substantial volume, *A Dissertation upon the Philosophy of Light, Heat, and Fire* (1794), criticizing Lavoisier's antiphlogistic chemistry; and, above all, a set of three massive quartos, *An Investigation of the Principles of Knowledge and the Progress of Reason, from Sense to Science and Philosophy* (1794), consisting of a wide-ranging discussion of human knowledge and its sources in sensation, perception, conception, passion, and action. See Dean, *James Hutton and the History of Geology*, 58–60.

106 Charles W. J. Withers, "On Georgics and Geology: James Hutton's 'Elements of Agriculture' and Agricultural Science in Eighteenth-Century Scotland," *Agricultural History Review* 42, no. 1 (1992): 38–48.

107 Bailey, *James Hutton*, 6–7.

108 Hutton, *Theory of the Earth*, 1:5.

ments of Agriculture.[109] Anticipating his geological writings, Hutton regarded the soil as a wasting asset, continually being washed away by rain and exhausted by the growth of plants and needing to be replenished as much by the slow breakdown of the underlying rocks as by the decay of plants, and held that the habitability of the land therefore depended on its slow disintegration. In the systematicity of the geological economy, the biological sciences appeared to Hutton to be completely integrated with the physical sciences: at this interface there opened up to him a compound system of things, forming together one organic whole precisely by way of an organization of forces interacting with one another. Applying what he had previously established in *An Investigation of the Principles of Knowledge,* Hutton proposed that agriculture represented a certain controlling power through which technology could cooperate with the means employed by nature to further the end of a fertile and habitable globe. In other words, humanity improves nature for the purpose of an increasingly ordered world. In the breeding of animals, he writes, the human merely "sets about improving those useful qualities which he finds naturally in the race."[110] In this way, there may be found in what Hutton calls "human art" a poietic disclosure of products in nature, products that are merely waiting to be born.

Such an aesthetic response to the self-organization of the earth generates the ambivalence of a landscape both sublime and economic.[111] Although Hutton appealed for evidence of these forces to aesthetic categories of sublime power and alien physicality, the implications of his theory are unmistakably similar to the Romantic notion that humanity, in recognizing itself as an element participating in nature's self-organization, will find its own freedom to be perfectly in line with the laws of nature. On

109 Jean Jones, "James Hutton's Agricultural Research and His Life as a Farmer," *Annals of Science* 42, no. 6 (1985): 574, 578–80.

110 James Hutton, *An Investigation of the Principles of Knowledge, and of the Progress of Reason, from Sense to Science and Philosophy,* vol. 2 (Edinburgh: Strahan and Cadell, 1794), 500.

111 Heringman, *Romantic Rocks, Aesthetic Geology,* 161.

the one hand, the Huttonian discourse on the rock record clearly worked to condition early industrial perceptions of human agency vis-à-vis the natural world: the sovereignty over nature that technology appeared to allow for was challenged by an aestheticization that reinserted the heat engine back into nature. On the other hand, however, the power afforded to humanity during the industrial revolution to manipulate and shape the natural world in ways previously unimaginable, and to master the forces of nature in such a manner as to rid it of its sublimity, prompted in the imagination a reciprocal interconnection between nature and artifice and spurred the need to understand humankind's artificial products in relation to the earth's natural products anew. No longer consigned to merely imitate the greatness of nature, this, in consequence, is what was required in order to make sense of the newfound technological power of modernity. Philosophers and naturalists alike had to dare to pursue nature into its most unlikely phenomena, leaving behind their comfort zones of primeval forests and unrestrained wilderness so as to go after those forces that sat less easily with their ethos. Humans had to rethink their relationship to the earth and to reconceptualize the sublime, which now seemed to have been banished from nature by their newfound technical prowess.

It is important to remember, however, that such a sublime production indicates but a novel form of a dynamically static harmony. In Hutton's geotheory, the becoming of the earth ultimately resulted in absolute stasis, which came to expression in his radically ahistorical and eternalist worldview, and which would later be rendered methodologically systematic, with the help of Whewell and Lyell, as the "Doctrine of Uniformity." This is made particularly evident in Hutton's model of total terrestrial fungibility by way of volcanic recontexture, whereby he imagined the formation of a future earth through strata currently being cooked within the mantle and then expelled at the bottom of the ocean.[112] For when considering the system of the earth, Hutton and his followers could only attend to what inci-

112 Hutton, "Theory of the Earth," 89.

dental evidence was left over for them to discover after centuries or even millennia of the rock's exposure to erosive forces had already destroyed the very proof required to support their theory of a dynamical equilibrium between deterioration and regeneration. Although they could point to the immense time span required to lay down the rock in stratified layers, they were in effect unable to derive from the hardened state of the strata any support for the fundamentally fluid process of recycling, which left them with an indeterminate number of such cycles, continuing without apparent end. As a result, the earth was imagined to be temporally becoming, but it was not, according to the Huttonian implications of deep time, becoming toward some definite end beyond its own endless perpetuation of itself. And although the objects in time—the natural and artificial products—were certainly finite, the becoming of the earth itself was nevertheless thought to be infinite. But precisely, therefore, were these future worlds recognized by Hutton only insofar as they were utterly identical with the present. And since the future was just the disinterment of the deepest past, Huttonianism in effect enforced a kind of preformationism of worlds. In this manner, Huttonian geology undoubtedly opened up the planet's physical depth. But by ungrounding it in an abyss of infinite productivity, Huttonian geology concomitantly flattened the ontological difference between world and earth.

3

Dissolving Technology, Planetary Metamorphosis

Two years before his death, the French philosopher and urban sociologist Henri Lefebvre published a short essay titled "Quand la ville se perd dans une métamorphose planétaire" (1989) — subsequently translated as "Dissolving City, Planetary Metamorphosis" — in which he contemplated the erosion of the ontological distinction of the artificial from that of the natural, which thus far had been essential to the meaning of urbanity. No longer did the city appear to him as a vehicle for the utopian realization of a different future. Rather, it was at once establishing itself as hegemonic and, paradoxically, in extending its urban fabric across the entire planet, disappearing into the unreflexive background of immediate environment. It is not only that global city formations and megacity expansions mark the imprint of the urban form onto larger and larger physical territories. The transition in question, he suggested, is not confined to the physical environment. Because alongside such a transformation to the socio-spatial arrangement of human settlements, we are also confronted with its successful colonization of the human imagination. Whereas nature used to exist outside the city limits, Lefebvre declared that, today, with "the planetarization of the urban,"[1] no corner of the world is beyond its instrumental organization of space. In place of the promise of modernization that was celebrated by architects such as Le Corbusier and Oscar Niemeyer, Lefebvre saw the threat of homogenization and the annihilation of diversity into the functionalist appropriation of the landscape of being, operationalizing social life solely in accordance with the means of ever-greater efficiency and utility. According to Lefebvre, the becoming-exclusive of expert knowledge and instrumental planning as the mode of being within which modern humans dwell is the essential feature of the emergent ontology of the planetary-technological fabric.

Of course, it is possible, by extending Lefebvre's observation, to speak not only of a dissolving city but of a dissolution of the artificial as a category more broadly. The novelty of Lefebvre's observation thus goes beyond the disciplinary boundaries of urban theory. In its more radical form, it concerns the significance of the spherical image of the earth, invoked, for instance, in the discourse of the contemporary debate about global environmental change. For if nature used to encircle humankind in a breathable, acclimatized, air-conditioned, and, in short, safe environment, as if within the walls of our urban conglomerates, then modernity arguably put us on a path toward an essentially artificial earth where technology was eventually nowhere to be found. Just as we install heating and ventilation in our buildings, and artificial regulation of temperature and humidity in our rooms, so there is a transition, in modernity, toward understanding nature more generally in terms of technical modification. In other words, there is a transition from understanding the atmosphere in terms of a ready-made, fixed, and transcendent condition to that of an immanent and processual "air design" — an "atmotechnics,"[2] to borrow from the vocabulary of the German philosopher Peter Sloterdijk — whereby, in order to

1 Henri Lefebvre, "Dissolving City, Planetary Metamorphosis," trans. Laurent Corroyer, Marianne Potvin, and Neil Brenner, *Environment and Planning D: Society and Space* 32, no. 2 (2004): 205.

2 Peter Sloterdijk, *Terror from the Air*, trans. Amy Patton and Steve Corcoran (Cambridge: MIT Press, 2002), 23.

persist in a certain state, it must be continually reproduced. As the anthropologist Bruno Latour put it, the properly modern question is, "What sort of materials constitute the walls that keep the *Dasein* from suffocating? In short, what is the climate in its air-conditioning system?"[3] As suggested by Sloterdijk and Latour, modern experience is in effect characterized by the notion that the spontaneous immunity services of humankind's terrestrial abode cannot be taken for granted anymore but will instead depend on human ingenuity and attention.

In the wake of modernity, then, the spherical image not only implies that humanity is enclosed within an environment, but also that humanity constructs and maintains the global immunity services as if from without. As each is the topological inversion of the other, this ambivalence consists in a fundamental ambiguity as to on which side of the circular contour that humans dwell: at one and the same time inside the confines of a sphere, bounded by the limits of its outer layer, and externally, on the surface of a globe, as if fundamentally responsible for the upkeep of its terrestrial home. Whereas in the first instance humankind is already restricted by the conditions of its being on the earth, in the second instance these conditions are produced by humankind, as if standing outside of itself looking in. In the words of the anthropologist Tim Ingold:

> To regard the world as a sphere is at once to render conceivable the possibility of its logical inverse, the globe; and of course vice versa. We could say that both perspectives are caught up in the dialectical interplay between engagement and detachment, between human beings' involvement in the world and their separation from it, which has been a feature of the entire history of Western thought and no doubt of other traditions as well.[4]

---

3 Bruno Latour, *Facing Gaia: Eight Lectures on the New Climatic Regime*, trans. Catherine Porter (Cambridge: Polity Press, 2017), 123.

4 Tim Ingold, "Globes and Spheres: The Topology of Environmentalism," in *Environmentalism: The View from Anthropology*, ed. Kay Milton (London: Routledge, 1993), 41–42.

Such is the ambivalence by which the introduction of the organism into the geological economy through the addition of a biotic sphere — a biosphere — ended up being by no means as one-sided as to provide intellectual support to the precautionary attitude often associated with the birth of environmentalism. As we shall see, this process of biospheric reproduction has also been interpreted as no less natural than the metamorphosis of the animal — the inner transformation of nature into novel forms of existence.

A Topology of Spheres

The sphere has been a master metaphor of the Western tradition ever since its Eleatic inception. From its very beginning with Parmenides, metaphysics proclaimed being to be "in a state of perfection from every viewpoint, like the volume of a spherical ball."[5] This is because spherical geometry, uniform in all directions from center to circumference and rotationally invariant, encodes exhaustive containment. Incidentally, this is also why the sphere offers itself as the default spatial format of ontological idealism. For if the container excludes qualities of the contained, then exhaustive inclusion and explanation cannot be achieved. Thus, spherical containers perfectly code for the epistemologically foundationalist commitments inherent to such idealist systems: reason is understood to be perfectly contained because the world that contains it is itself inherently reasonable. As Sloterdijk has noted, such an amniotic inclusion of reason within being remained the fundamental metaphoric function of spheres until the eighteenth century, which he aptly diagnoses as "the twilight of the orb epoch," historically marking the "collapse of the metaphysical immune system"[6] that was once prof-

5 Allan H. Coxon, *The Fragments of Parmenides: A Critical Text with Introduction and Translation, the Ancient Testimonia and a Commentary,* trans. Richard McKirahan (Las Vegas: Parmenides Publishing, 2009), 78.

6 Peter Sloterdijk, *Globes — Spheres,* vol. 2: *Macrosphereology*, trans. Wieland Hoban (Los Angeles: Semiotext(e), 2014), 43–45, 559.

fered by the geometric "inclusion figure[s]"[7] of Greek antiquity. In the wake of Nicolaus Copernicus's shattering of the Ptolemaic spheres, humanity, for the first time, found itself in a state of "shell-lessness."[8] For if the "mental atmosphere" of antiquity was artificially air-conditioned by the exhaustive containment of the immunological sphere, then its inclusivism functioned precisely by concealing the artificial limits to its enclosure. Thus, blowing a hole in the celestial sphere to reveal the eternal silence of an infinite space beyond the inherently meaningful cosmos of the ancient Greeks not only exploded the limits of its inclusion figures from within, but concomitantly revealed the very limitations of inclusivism as exhaustive containment.

Well before Immanuel Kant's First Critique, then, the spherical shape of the geocosm no longer communicated irreducible inclusion of dwellers within their dwelling. Admittedly, the guiding idea of the ordered whole could still be found in its modernized form in the Kantian idea of the phenomenal world, that is, order as a kind of transcendental projection — Kant even argued that "reason is not like an indeterminably extended plane, [...] but must rather be compared with a sphere."[9] In accordance with this analogy, the topology of the earth's surface stands in for the fundamental idea — which the mind is said to bring to experience — of the unity, systematicity, and uninterrupted continuity of the globe upon which the spirit of our phenomenal experience is doomed to roam. Relegating the sphere metaphor to the domain of mind, teleology was, after Kant, a demand of reason rather than a naked fact of nature. But even though a critique of pure reason entails that we ought no longer to dogmatically accept as a given that what exists forms a coherent unity, it is nevertheless so that to know something still requires us presuming that it fits into a system of knowledge.

---

7 Peter Sloterdijk, *Bubbles — Spheres*, vol. 1: *Microspherology*, trans. Wieland Hoban (Los Angeles: Semiotext(e), 2011), 329.

8 Ibid., 24.

9 Immanuel Kant, *Critique of Pure Reason*, ed. and trans. Paul Guyer and Allen W. Wood (Cambridge: Cambridge University Press, 1998), 654–55 (A762–A763/B790–B791).

Hence, we must continue to act and reason as if there is such a systematicity to nature, for without the transcendental notion of a world picture — like a mental version of the Apollo 17 snapshot of the earth from space — there would be no organization of the bits and pieces of our cognitions into the sensible structure of a frame. The idea of the globe precedes our cognitions of the terrestrial environment, but it contains within itself the necessary conditions that make it possible to determine the place and relation of all partial cognitions so as to shape it into the smooth and continuous experience that characterizes the spherical shape of the phenomenal world.[10] Though the cascading content of sensation is potentially infinite — in the sense that traversing a sphere's continuous surface provides no boundary — the space of reasons governing it, that is, the discursive rules and functional norms regulating sense experience, nonetheless generates bounds or limits, just as do the contours of the spatially finite sphere.[11] In contradistinction to the geometric inclusion figures of the ancient Greeks, Kant's sphere metaphor, therefore, served not to ground reason's warrants within brute existence through some including and inclusive foundation, but instead expressed our reliance of such warrants on values of holism and coherence that are spontaneous and exclusive to the discursive intellect.

As Kant reserved place within his idealism for an essential interiority, he also put the modern concern for limits at center stage. Here, then, the metaphorical sphere serves a new and distinct expressive purpose: not to reiterate the submission of cognition to the tyranny of some foundational existence, but instead to consummate that aspect of cognition whereby it can become self-governing and thus legislate its contents relative to a global criterion of correctness that supersedes and opposes the merely parochial and exigent authorities of sense data, received doxa, or commonsense intuitions. Yet, in precise concomitance

10 Marie-Eve Morin, "Cohabitating in the Globalised World: Peter Sloterdijk's Global Foams and Bruno Latour's Cosmopolitics," *Environment and Planning D: Society and Space* 27, no. 1 (2009): 63.

11 Kant, *Critique of Pure Reason*, 653–55 (A759–A762/B787–B790).

with its expatriation from such ontic foundations, reason reneges the specifically ontic insurance, enclosure, and protection that spheres once provided.[12] Contrary to the inclusivism of the geosomatic order (where the city-state is ideally organized in accordance with the harmonious coexistence of planetary bodies in the cosmos, since the cosmos is but the model of an ideal city-state whose perfection is guaranteed by divine law) that dominated the imagination of the ancient Greeks, the core contention of the Kantian sphere is that autonomy goes hand in hand with self-restriction, such that responsibility always entails risk. Thus the feminist science scholar Donna Haraway could later declare that the immune system is an iconically mythic object that functions by drawing out boundaries "to guide recognition and misrecognition of self and other in the dialectics of Western biopolitics."[13] In the wake of modernity, inclusion presupposes exclusion. As we transition, in Sloterdijk's terms, from the immunological security of the sphere to the spherological precarity of immunity, we are no longer merely born into a passive defense system enclosing the earth in spherical forms like heavenly mantles. Securing ourselves against threat instead becomes an active project of collective immune design, in the process of which the distinction between inside and outside is produced rather than a given.

Notwithstanding this, Kant noted of his comparison between discursive and planetary spheres—metaphorically appropriating the notion of the earth's hidden depths—that we are "ignorant in regard to the objects that this surface might contain."[14] Indeed, the Kantian traveler, for whom the world is a globe, is forsaken to journey upon the outer surface of the earth. For it is at this surface—the interface between world and mind, sensation and cognition—that all knowledge is constituted. There

12 Thomas Moynihan, *Spinal Catastrophism: A Secret History* (Falmouth: Urbanomic, 2019), 17–20.

13 Donna J. Haraway, "The Biopolitics of Postmodern Bodies: Constitutions of Self in Immune System Discourse," in *Simians, Cyborgs, and Women: The Reinvention of Nature* (London: Routledge, 1991), 204.

14 Kant, *Critique of Pure Reason*, 653 (A759/B787).

is, consequently, no naïve going back, in the wake of Kant, to humankind's immediacy in nature, that is, to its dwelling within the inclusivism of the antique cosmos.[15] The globe does, of course, still have a core, but the very kernel of the thing-in-itself is forever barred to human access. Beyond humankind's dwelling upon the terrestrial surface lies a fundamentally unknown domain of sublime, primordial forces inaccessible to representation and calculation. However, when exploiting this sentiment, Arthur Schopenhauer later wrote that "consciousness is the mere surface of our mind, and of this, as of the [terrestrial] globe, we do not know the interior, but only the crust,"[16] and so we can understand his rejoinder to Kant as a reiteration of the notion of the inorganic unconscious inaugurated by the post-Kantian insistence on the antecedence of a chthonic irrationalism — recall the German Romantics' adoption of Huttonian deep time for ontological means, so as to return reason to its place within nonconceptual nature by way of establishing a depth-psychological dimension to the noumenal realm, thus reconnecting modern humans to nature again.[17] With humankind's decentering from the heart of the modern geocosm, and with the primacy of reason under threat by the inorganic antecedence of an inhospitable and uncaring nature, this desire for systematicity, unsatisfactorily addressed by Kant's regulative compromise, could find its resolution only in an inverted form, that is, through an ontological univocity that rendered artifice natural precisely by disclosing nature as essentially artificial. For although the metaphysical inclusivism that grounded the ancient world was fundamentally threatened by the Copernican penetration of the immunological boundaries of the antique spheres, it does not mean that the sphere was consigned to the dustbin of history altogether. On the contrary, in place of the metaphysical immunology that secured the inclusivism of the

15 Ingold, "Globes and Spheres," 36–37.
16 Arthur Schopenhauer, *The World as Will and Representation,* trans. Eric F. J. Payne, 2 vols. (New York: Dover, 1969), 2:136.
17 Moynihan, *Spinal Catastrophism,* 25–29.

ancient Greek geocosm, a number of geophysical Ersatz-spheres quickly stepped in to fill the void that the former had left in the cultural imagination of a now shell-less civilization.

§

A crucial step in this direction was taken in 1875, when Eduard Suess, an Austrian geologist and expert on the Alps at the University of Vienna, first utilized the term "biosphere" to designate the zone of the earth's surface that contains biological lifeforms. Even though similar but more expansive concepts employing the idea of a biotic enclosure that covered the earth's crust had already been proposed by other nineteenth-century scholars, such as the German geographer Ferdinand von Richthofen,[18] most historians and scientists have retroactively credited Suess as its founder.[19] In his treatise on the genesis of mountains, *Die Entstehung der Alpen* (1875), Suess laid claim to the term as follows:

> One thing seems to be foreign on this large celestial body consisting of spheres, namely, organic life. But this life is limited to a determined zone at the surface of the lithosphere. The plant, whose deep roots plunge into the soil to feed, and which at the same time rises into the air to breathe, is a good illustration of organic life in the region of interaction between the upper sphere and the lithosphere, and on the

---

18 Jonathan D. Oldfield and Denis J. B. Shaw, "A Russian Geographical Tradition? The Contested Canon of Russian and Soviet Geography, 1884–1953," *Journal of Historical Geography* 49 (2015): 79.

19 G. Evelyn Hutchinson, "The Biosphere," *Scientific American* 223, no. 3 (1970): 45; Lynn Margulis and Dorion Sagan, *What Is Life?* (Berkeley: University of California Press, 1995), 50; Nicholas Polunin and Jacques Grinevald, "Vernadsky and Biospheral Ecology," *Environmental Conservation* 15, no. 2 (1988): 118; Václav Smil, *The Earth's Biosphere: Evolution, Dynamics, and Change* (Cambridge: MIT Press, 2002), 1–2; and James E. Lovelock, *The Revenge of Gaia: Why the Earth Is Fighting Back and How We Can Still Save Humanity* (Harmondsworth: Penguin, 2007), 160.

surface of continents it is possible to single out an independent biosphere.[20]

Clearly, the concept of the biosphere, right from the onset, contained physical and metaphysical implications. To begin with, the biosphere, for Suess, indicated the living sphere of the earth as a network of individual organisms, which, in a manner similar to the geophysiology of Hutton, were depicted as parts belonging to a greater whole. However, Suess used the term only in passing and without further elaboration, leaving behind a strikingly limited definition. Unsurprisingly so, because his intention was only to make a distinction between the lithosphere, making up the crust of the earth's upper mantle, and the thin film on top of it, where living beings dwell. But though explicable in terms of his professional preoccupation with crystal forms and orogenesis, Suess's definition of the biosphere still revealed a glaring exclusion of marine life, nor did he consider microorganisms, which are abundant both in the lower layer of the atmosphere and in the ocean. And, what is more, in conceptualizing the biosphere as Selbständig, he seemed to deny the myriad links between organisms and the rest of the earth's geospheres.

As a result, the term "biosphere" was slow to find common use and would not catch on within the scientific community until at least twenty-five years later, when it was picked up by the Russian mineralogist and geochemist Vladimir Vernadsky. One reason for this was that Vernadsky clarified the meaning of the concept through a detailed examination of its internal structure

20 Smil, *The Earth's Biosphere*, 1. This translation from the German is by Václav Smil. In the original passage, it reads as follows: "Eines scheint fremdartig auf diesem grossen, aus Sphären gebildeten Himmelskörper, nämlich das organische Leben. Aber auch dieses ist auf eine bestimmte Zone beschränkt, auf die Oberfläche der Lithosphäre. Die Pflanze, welche ihre Wurzeln Nahrung suchend in den Boden senkt und gleichzeitig sich athmend in die Luft erhebt, ist ein gutes Bild der Stellung organischen Lebens in der Region der Wechselwirkung der oberen Sphären und der Lithosphäre, und es lässt sich auf der Oberfläche des Festen eine selbständige Biosphäre unterscheiden"; Eduard Suess, *Die Entstehung der Alpen* (Vienna: W. Braunmiller, 1875), 159.

and features. Following Suess, Vernadsky took the biosphere to constitute the enclosure of the earth's crust by an envelope of organisms. But he also diverged from Suess's definition by arguing that the biosphere ought to be understood in relation to the elements of inorganic nature that afford organisms the necessary conditions for their existence. The products of organisms' activities on their environment, such as photosynthesis and oxygenation, cannot be taken as independent from the elements and chemical compounds involved in these vital processes of transformation. Crucial for the possibility of carbon-based life, water too must be considered in relation to the biosphere, as must solar radiation, which is equally fundamental for the maintenance of life on earth. According to Vernadsky, then, it is at the interface between organic and inorganic nature that we find the conceptual significance of the biosphere: it constitutes the ambivalent intersection between the organic and the inorganic, where life is engaged in producing the habitat for its own existence, which brings to mind the same kind of well-managed estate that had served as a suitable metaphor for Hutton more than a century before. Not only are organisms conditioned by their environment, but the environment itself, Vernadsky pointed out, is always produced by organisms, and humankind in particular, Vernadsky increasingly stressed toward the end of his career, constitutes a remarkable force of change in the composition of the biosphere. In line with his refinements, the biosphere was later defined by G.E. Hutchinson, one of the leading ecologists of the twentieth century, as that region of the earth's geological economy within which organisms dwell, but with several qualifications as to the diffusion of the vital processes both upward, into the earth's atmosphere, and downward, into its crust.[21] Building further on the work of Hutchinson, the botanist Nicholas Polunin together with the intellectual historian Jacques Grinevald sought in the 1980s to offer a more precise redefinition of the biosphere, by going back to Vernadsky, as the "integrated living and life-supporting system comprising

21 Hutchinson, "The Biosphere."

the peripheral envelope of Planet Earth together with its surrounding atmosphere so far down, and up, as any form of life exists naturally."[22]

The first comprehensive treatment of the biosphere by Vernadsky was published in Russian in 1926, under the title *Biosfera,* but it first brought him international attention three years later when it was translated into French.[23] The work itself was separated into two sections, both of which sought to establish the biosphere as continuous in many ways with factors both external (powered by the sun) and internal (interacting, for instance, with the earth's atmosphere, hydrosphere, pedosphere, and lithosphere) to the planet. Key to Vernadsky's conceptual contribution was his contention that "the biosphere may be regarded as a *region of transformers* that converts cosmic radiations into active energy in electrical, chemical, mechanical, thermal and other forms."[24] His contribution was thus to underline how the transformation of free energy into material production by organic processes bestows upon the earth an

22 Polunin and Grinevald, "Vernadsky and Biospheral Ecology," 118.

23 Although it has since been translated into several languages, *The Biosphere* did not appear in English until 1986. Much of Vernadsky's early work first appeared in French scientific journals, but most of his major works, including his last and unfinished magnum opus, "The Chemical Structure of the Biosphere and Its Surroundings," exist only in Russian. For many years, the only pieces of Vernadsky's writing available in English were two articles published in 1944 and 1945 — "Problems in Biochemistry, II: The Fundamental Matter-Energy Difference between the Living and the Inert Natural Bodies of the Biosphere" and "The Biosphere and the Noösphere," which both served as concise summaries of his theories. They were translated by the author's son, George Vernadsky, at the time a historian at Yale University, with the help of G.E. Hutchinson, then a young limnologist, who took a great interest in Vernadsky's attention to the relationship between biological communities and the nonliving components of their environment, and who would later go on to apply Vernadsky's ideas to the study of biogeochemical processes taking place in a small self-contained ecological system, Linsley Pond, near New Haven, Connecticut. See Mercè Piqueras, "Meeting the Biospheres: On the Translations of Vernadsky's Work," *International Microbiology* 1, no. 2 (1998): 165–66.

24 Vladimir I. Vernadsky, *The Biosphere,* trans. David B. Langmuir (New York: Copernicus, 1998), 47 (my emphasis).

extraordinarily self-organizing character.[25] Given the importance of the continual exchange of matter not only within the geological economy but also between the earth and its cosmic environment, Vernadsky even speculated that life may initially have been brought to our planet's surface from elsewhere in the universe, only to begin terraforming the earth once the conditions for such a process to take place were present. Foreshadowing the methodological approach that would come to fruition in the space programs of the latter half of the twentieth century two decades after his death, Vernadsky urged the examination of extraterrestrial objects to determine their chemical composition and insisted on the study of the chemistry of other planets in order to compare their (dis)similarities to that of the earth. As we shall see in the concluding chapter, this proposal, and the conceptual framework that underlies it, bears obvious similarities with the theory put forward by James Lovelock and Lynn Margulis almost fifty years later, but, according to their own testimonies, they were both unfamiliar with Vernadsky's work at the time of their writing.[26]

Vernadsky credited Suess with coining the term, but he considered the idea of a biosphere ultimately to be derived from the French naturalist Jean-Baptiste de Lamarck, who had argued as early as 1802, with the publication of his *Hydrogeology*, that a systematic terrestrial physics must include not only geology but also meteorology and biology.[27] In fact, the novelty of Vernadsky's conceptual development of the biosphere was his strikingly organicist bent, with its focus on the self-organization of the earth taking place at the interface between organism and environment. At least in part, his sensitivity to the continuity between the organic and the inorganic can be attributed to

25 Ibid., 44.

26 James E. Lovelock, "Prehistory of Gaia," *New Scientist* 111 (1986): 51; Lynn Margulis, "Foreword to the English-language Edition," in Vernadsky, *The Biosphere*, 16; and Andrei G. Lapenis, "Directed Evolution of the Biosphere: Biogeochemical Selection of Gaia," *The Professional Geographer* 54, no. 3 (2002): 383.

27 Hutchinson, "The Biosphere," 45.

the academic setting within which he came to maturation as a young scientist. During the early, formative part of his career, Vernadsky trained under the supervision of the chemists Alexander Butlerov and Dimitry Mendeleev, and the pedologist Vasily Dokuchaev.[28] Above all, Dokuchaev's transdisciplinary approach to soil morphology as a global natural object at the intersection between geology, mineralogy, meteorology, chemistry, and biology — again, not at all unlike Hutton's analogy of the functioning of the earth to that of a well-run farm — was foundational for Vernadsky's later work to establish biogeochemistry as a new field of scientific enquiry.[29] In his youth, as he accompanied Dokuchaev on expeditions to study the black soil of the Eurasian Steppe, Vernadsky's attention was attracted to the organic processes that actively contributed to the fertility of the earth. From Dokuchaev's methodological point of view, the study of the pedosphere could not be reduced to investigating its geophysical and chemical composition. Rather, it had to be approached from the point of view of its role in the evolutionary history of microbial life too, that is, as being acted upon, across time, by a variety of microorganisms, plants, and animals. Vernadsky's early work on soil formation and conservation certainly constituted an important stimulus for developing a material understanding of the earth as already characterized by certain vital tendencies rather than being inert and static, acknowledging not only its geological, mineral, and climatic but also its biotic and anthropogenic factors.[30]

If anything, the lasting effect of this formative experience was to nudge him in the direction of a deep-seated doubt that the structure and composition of the earth could ever be fully explained with reference to abiotic processes alone. "I realized,"

28 Kendall E. Bailes, *Science and Russian Culture in an Age of Revolutions: V.I. Vernadsky and His Scientific School, 1863–1945* (Bloomington: Indiana University Press, 1990), 17–18.

29 Ibid., 37, 186–87.

30 Andrei V. Lapo, *Traces of Bygone Biospheres* (Santa Fe: Synergetic Press, 1998), 28, and Bailes, *Science and Russian Culture in an Age of Revolutions*, 18–21.

Vernadsky would later recall, "that the basis of geology lies in the chemical element — in the atom — and that living organisms play a prominent role, perhaps the leading one, in our natural environment — the biosphere."[31] In an effort to reintegrate the biotic with the abiotic, Vernadsky drew a crucial distinction between "life" and "living matter." "'Living matter,'" he wrote, "is the totality of living organisms. It is but a scientific empirical generalization of empirically indisputable facts known to all, observable easily and with precision. The concept of 'life' [on the other hand], always steps outside the boundaries of the concept of 'living matter'; it enters the realm of philosophy, folklore, religion, and the arts."[32] Although he perceived the former as an emergent phenomenon from biogeochemical processes in the natural world, Vernadsky considered the concept of life too bound up with an assumed separation between, on the one hand, a mechanistic universe and, on the other, the autonomy of organisms.[33] Such a separation seemed to him increasingly untenable and obstructive in the light of empirical investigations. From his point of view, Cartesianism had largely turned out to be powerless to compensate for the self-conscious unity connecting humanity to the earth processes and thus lagged behind the demands of the natural sciences. In this sense, Vernadsky's terminology can be understood as a retort against the kind of ontological dualism held by many of the contemporary vitalists. But it was also directed as a response to the one-dimensional framework of mechanistic philosophy, wherein spirit had been outright eliminated, or, at best, been reduced to a perplexing anomaly in an otherwise purposeless world consisting of nothing but efficient causes. Instead, at that liminal interface between living and inert matter that Suess had only

31 Vernadsky, quoted in Bailes, *Science and Russian Culture in an Age of Revolutions*, 185.

32 Vladimir I. Vernadsky, "The Biosphere and the Noösphere," *American Scientist* 33, no. 1 (1945): 6.

33 Jonathan D. Oldfield and Denis J. B. Shaw, "V.I. Vernadsky and the Noosphere Concept: Russian Understandings of Society–Nature Interaction," *Geoforum* 37, no. 1 (2006): 149.

begun to outline in terms of a biotic sphere, Vernadsky saw one single cycling of materials and energy that did not care much to conform to religious or philosophical boundary work. Such a biogenic migration of chemical elements from the biotic to the abiotic and back again, maintained only through the processes of respiration, alimentation, and reproduction of living matter, yet absolutely crucial for the continued survival of the biota, seemingly blurred the distinction between environmental determinant and voluntaristic enactment. If the concept of the biosphere in the work of Suess, following Kant, further paved the way for the transition from a natural trinity (mineral, vegetal, and animal) to a binary (organic and inorganic), the biosphere as developed by Vernadsky aspired, on the contrary, to reinstate the unity of nature in the Romantic tradition.[34]

For Vernadsky, living matter — understood as a vital expression of material processes — constitutes the most fundamental force in the evolution of the earth, which means that the study of geochemistry must be realigned into a *bio*geochemisty. At the same time, then, what Vernadsky understood by "living" was something much more comprehensive than the commonly received definitions of the contemporary biosciences. Life *is* biospheric: it encompasses the ensemble of biota inhabiting the planet, from the microbial to the human. Furthermore, it is self-regulating and self-responsive. Far from adapting to external geochemical conditions, the evolution of any one lifeform is in the first instance determined by its relationships to other forms of life. Importantly, Vernadsky did not believe that life was evenly distributed across the universe, nor did he think that it was "an external and accidental phenomenon of the Earth's crust." On the contrary, "it is closely bound to the structure of the crust, forms part of its mechanism, and fulfills functions of prime importance to the existence of this mechanism," and thus so fundamental to the peculiarly hospitable environment of our planet that without life, "the face of the Earth would become as

34 Bertrand Guillaume, "Vernadsky's Philosophical Legacy: A Perspective from the Anthropocene," *The Anthropocene Review* 1, no. 2 (2014): 139.

immobile and chemically passive as that of the moon."[35] Vernadsky thus insisted that the discipline of biogeochemistry was concerned with metabiotic processes rather than with life as such, and with the planetary impact of "living matter as a whole — the totality of living organisms"[36] as opposed to the role of separate taxas. For considered on a biospheric scale, the defining characteristic of the totality of organisms is its metabolic transformation of nature through the constant supplying of matter, shaping it into new forms by altering its structure.[37] The vital tendency of living matter is to co-opt its environment, such that the most important geological phenomenon is precisely that which, with the withering away of natural philosophy into the disciplinarity of the specialized sciences, geologists had surrendered to be studied exclusively by biologists. "It is evident that if life were to cease the great chemical processes connected with it would disappear," Vernadsky declared, "both from the biosphere and probably also from the crust."[38] Here, the biotic serves as but an extension of the abiotic.

By positioning living matter in the midst of other forces in the earth's geological economy, Vernadsky committed heresy on two counts — two heresies melded into an ontologically symmetrical scale containing the biotic and the abiotic: geosomatic vitality as ever-renewing mineral, and, inversely, minerals as but the slowest renewing and least complex expression of geosomatic vitality.[39] First, he challenged biologists by considering the biosphere of living organisms as a large chemical factory. Plants and animals were mere vehicles for the massive flow of elements constituting the earth's geological economy, because from the point of view of its geochemical implications,

35 Vernadsky, *The Biosphere*, 57–58.

36 Ibid., 58.

37 Ibid., 50–60.

38 Ibid., 56.

39 Kevin Kelly, *Out of Control: The New Biology of Machines, Social Systems, and the Economic World* (Reading: Perseus Press, 1994), 75–77.

> it is necessary to present [living matter] in the same terms, with the same logical parameters, as other forms of existence of chemical elements to which we are comparing it here; that is, minerals, rocks, magmas, water solutions, and dispersions. In other words, the totality of organisms must be expressed only from the standpoint of their mass, their chemical composition, their energy, their volume, and the character of the space corresponding to them.[40]

Living matter was, ontologically speaking, like a specific kind of rock — an ancient and, at the same time, eternally young rock, because "the animal realm," according to Vernadsky, "does not manifest life in itself."[41] On the contrary, seen from a geochemical perspective, the function of animals and plants is rather to assist the wind and waves to stir the brewing biosphere, for taken as a whole, living matter manifests itself as an increase of active energy in the earth's crust. Second, then, he also challenged geologists by considering rocks as if they were but the slowest moving matter in the nonlinear causality of the planetary body's self-organization. The mountains of the lithosphere, the waters of the hydrosphere, and the gases of the troposphere were by no means inert, as if existing in a stillborn equilibrium, but rather in a constant dynamic relationship with each other. Since Vernadsky found the same to be true of the biosphere's interaction with the other geospheres of the earth, the genesis of every rock was essentially in this nonlinear perpetuation of organic destruction and reparation.

However, such an observation was not entirely novel at the time. For as we saw in chapter 2, it is fundamentally Huttonian in nature. Furthermore, in the wake of the natural historical abyss of deep time, it was intuitively understood in more detail by evolutionary biologists already in the eighteenth century: the

40 Vladimir I. Vernadsky, *Geochemistry and the Biosphere: Essays by Vladimir I. Vernadsky*, ed. Frank B. Salisbury, trans. Olga Barash (Santa Fe: Synergetic Press, 2007), 77.

41 Ibid., 155.

environment shapes the organism, and the organism, in turn, shapes its environment. Anticipating Lovelock and Margulis's Gaia hypothesis, Vernadsky posited that the biosphere is not simply an accidental evolutionary twist in the history of the earth's geological development, but rather an expected and necessary emergence. Without organisms, the crustal mechanism of the earth would not be what it is.[42] And the same holds true moving down from the gaseous layers of earth's atmosphere toward the crust, for, as he puts it, "living matter can be taken as an appendage of the atmosphere."[43] This effort to fold life back into its terrestrial environment is essential for understanding Vernadsky's reflections on the essence of technology, stemming from his observation that there is a gradual increase, over time, in the number and kinds of chemical compounds entering into the geological economy, and also an incremental acceleration to the pace of their cycling.[44] Human artifice, for him, was thus merely an instantiation of this natural historical gain in momentum:

> Within the last five to seven thousand years the continuous creation of the Noösphere has proceeded apace, ever increasing in tempo, and [...] the increase of the cultural biogeochemical energy of mankind is advancing steadily without fundamental regression, albeit with interruptions continually diminishing in duration. There is a growing understanding that this increase has no insurmountable limits, that it is an elemental geological process.[45]

---

42 Vernadsky, *The Biosphere*, 58.

43 Ibid., 87.

44 Vladimir I. Vernadsky, "'Problems of Biogeochemistry' II: The Fundamental Matter-Energy Difference between the Living and Inert Natural Bodies of the Biosphere," ed. G. Evelyn Hutchinson, trans. George Vernadsky, *Transactions of the Connecticut Academy of Arts and Sciences* 35 (1944): 483–517.

45 Vladimir I. Vernadsky, "The Transition from the Biosphere to the Noösphere—Excerpts from Scientific Thought as a Planetary Phenomenon, 1938," trans. William Jones, *Twenty-First Century Science & Technology* (Spring–Summer 2012): 27–28.

Progressing along the evolutionary trajectory, the lithosphere precedes the biosphere, but the biosphere, as it comes into being, does not merely abrogate the lithosphere. Rather, both begin to take new shapes in relation to each other, and for Vernadsky the same holds true for the entry of human artifice—or, what he called "the noösphere," as is evident in the above quote—into the biosphere.

§

If Vernadsky laid the ground for the ensuing discussion about the artificialization of the earth, the technogenesis of a seemingly directed, natural evolutional tendency toward an intensified artificial transformation of our planet was, however, most vividly described by the Jesuit paleontologist Pierre Teilhard de Chardin, who, together with Édouard Le Roy, the disciple and successor of Henri Bergson at the Collège de France, attended Vernadsky's lectures on biogeochemisty at the Sorbonne in 1922.[46] Alert to the increasing reach and pace of technological development, and influenced by the evolutionary theories of Lamarck and Charles Darwin, Teilhard de Chardin was concerned with the sense in which the modern subject, beyond its use of technology, had also to adapt to these new machines, acting within a technological environment where the relationship between humans and their tools was becoming much more of an *inter*action than the one-way street of instrumental manipulation would imply. Because of the coevolution of organism and environment, it stands to reason, according to Teilhard de Chardin, that humankind's aptitude as a toolmaker has made it into the vehicle for the actualization of novel forms of artificial life. Importantly, however, Teilhard de Chardin understood this process as fundamentally symbiotic. Like the organism, the human is not only involved in shaping its environment technically but is destined to be shaped by the same technical infrastructures in a transformative symbiosis by which it, because

46 Bailes, *Science and Russian Culture in an Age of Revolutions*, 162.

of the alteration of its own nature, will gradually become other than what it currently is. Along with the increase in technological reach and pace, Teilhard de Chardin thought that it would become more and more evident that, rather than being defined by the singular essence of a preformed being, the human, just like the animal post-Darwin, is better understood as the function of distributed networks of constructed environments, infrastructures, and institutions. In this view, the emergence of technology embodied as radical a paradigmatic shift in natural evolution as the appearance of life and came to represent a stage in the actualization of human freedom as the consequence of nature's inner development. Gradually, as nature becomes increasingly self-conscious, the whole earth is enrolled in the universal process whereby the spiritual is eventually reconciled with the material. This, again, puts humankind in a fundamentally split position vis-à-vis the natural world: simultaneously over and above the deterministic mechanism of physical processes, and yet fundamentally a product of—and thus guided by—natural evolution.

Born out of his duality as both a paleontologist and a Jesuit priest, Teilhard de Chardin's entire philosophy features an interesting tension between reason and myth. As a Jesuit, the starting point for Teilhard de Chardin's writings was how to articulate the importance and uniqueness of Christianity for how humans can understand the natural world they live in, and for how they can understand their own place within it. However, because of his own experience as a practicing paleontologist, which allowed him to share in the existential anxieties of modern science, Teilhard de Chardin's faith was markedly different from that of many of his peers. Like Thomas Aquinas, Georg W. F. Hegel, and many other Christian natural philosophers before him, Teilhard de Chardin's task was one of bringing together the Bible with the Book of Nature by positing the role of the divine not in spite of what the sciences had discovered but rather *because of* their findings, where the question of the spiritual, instead of being external to and separate from the worldly concerns of the sciences, was conceived of as immanent and forever relevant to the world we seek to understand.[47] More specifically, Teilhard de Chardin championed "orthogenesis," particularly as it figured in the work of Bergson, which ran counter to what most biologists had insisted, and in fact have insisted ever since. Orthogenesis proposes that evolution does not stumble along randomly but rather constitutes an innate tendency in nature — teleologically indeed — to organize itself in increasingly complex forms. In this manner, Teilhard de Chardin speculated that the earth as such was involved in the ascent toward unleashing the inherent potentiality of spirit within matter, which would culminate in the realization that, echoing the Romantics, we as humans are participating in the delivery of spirit out of nature. Although humans certainly do not have a monopoly on consciousness, they nevertheless have, as a result of their inquiry into nature, started a self-reflective process that turns on the question of their own humanity. According to Teilhard de Chardin, humankind's location in the tree of life, albeit merely one element among many, is exceptional. As humans, we constitute a threshold in the reconciliation of matter and spirit, and we are teleologically drawn toward this goal.

"Of old," writes Teilhard de Chardin, "forerunners of our chemists strove to find the philosophers' stone. Our ambition has grown since then. It is no longer to make gold but life."[48] It is no longer the promises of alchemy — of turning lead into gold — that stands in as the ultimate object of desire of the sciences, even though the symbolism in this case is not at all inaccurate. Rather, bioengineering, the control of lifelike processes, is now what drives "the direction of a conquest of matter put to the service of mind."[49] Turning the whole evolutionary edifice upon its head, Teilhard de Chardin speculated that if the earth at first may have seemed to be made only of metals and minerals, this was only the most primitive form of a germinating seed

---

47 Pierre Teilhard de Chardin, *The Heart of Matter,* trans. René Hague (New York: Harcourt, 1979), 212.

48 Pierre Teilhard de Chardin, *The Phenomenon of Man,* trans. Bernard Wall (New York: HarperCollins, 2008), 249.

49 Ibid.

still in the initial stages of transmutation. Rather than depicting life as the happy accident of blind mechanism, he described the juvenile earth as already suffused with a vital potency, and added that "its activities, hitherto dormant, are now set in motion *pari passu* with the awakening of the forces of synthesis enclosed in matter [...] [as] over the whole surface of the new-formed globe, the tension of internal freedoms begins to rise."[50] The history of the earth, in other words, was viewed as merely the actualization of what was already present. In this sense, as explained by the French biochemist and geneticist Jacques Monod, "there is no 'inanimate' matter, and therefore no essential distinction between 'life' and 'matter.'"[51] First, the self-organization for and by the earth itself gives rise to subtle bodies of varying degrees of intensity, from the unconscious bodies of rock formations and plants, to the conscious bodies of animals, and lastly transforming itself into the self-conscious bodies of humans to become a vessel for the explicit expression of that which is implicit in nature, namely, the purposeful strive of the spirit. Of course, this is an esoteric representation of the movement toward more complex products of nature. But at the same time, such a symbolism does nothing to undermine the concurrent paleontological account of the phylogenetic tree that Teilhard de Chardin also provides. For alongside his more speculative inclinations, he meticulously recounts the history of life on earth, from self-replicating molecules on to unicellular organisms, eukaryotes, multicellular organisms, arthropods, mammals, primates, hominini, archaic humans, and, finally, into the self-reflexive beings that are anatomically modern humans—and "finally" should be understood literally in this case, because unlike most evolutionary theorists in the history of science, "the 'galaxy' of living forms constitutes," from Teilhard de Chardin's point of view, "a vast 'orthogenetic' movement of involution on an ever-greater

50 Ibid., 73.

51 Jacques Monod, *Chance and Necessity: An Essay on the Natural Philosophy of Modern Biology*, trans. Austryn Wainhouse (New York: Vintage Books, 1972), 31–32.

complexity and consciousness,"[52] such that it consists in a universal process tending toward that self-directed end.

Although Teilhard de Chardin was an accomplished paleontologist who assisted in the discovery of the fossil specimens of both the "Peking Man" and the "Java Man," his speculative style of writing, combined with his theological convictions, meant that his work for the most part received a harsh response from the scientific community. Many a biologist outright dismissed his evolutionary musings. Most notably, the 1960 Nobel Prize–winning immunologist and zoologist Peter Medawar disapprovingly wrote that Teilhard de Chardin's magnum opus, *The Phenomenon of Man* (1955), "for the greater part [...] is nonsense, tricked out with a variety of metaphysical conceits, and its author can be excused of dishonesty only on the grounds that before deceiving others he has taken great pains to deceive himself."[53] Another heavyweight name, Richard Dawkins, sided with Medawar, declaring that it was "the quintessence of bad poetic science."[54] However, other scholars of the biological sciences, no less distinguished, were far more sympathetic. For instance, Julian Huxley, popularizer of the "modern synthesis," which reconciled Darwinian evolution with Mendelian inheritance, provided an introduction to the English translation in which he concluded that, after Teilhard de Chardin, "the religiously-minded can no longer turn their backs upon the natural world, or seek escape from its imperfections in a supernatural world; nor can the materialistically-minded deny importance to spiritual experience and religious feeling."[55] Theodosius Dobzhansky, another towering figure in twentieth-century genetics, whose famous essay "Nothing in Biology Makes Sense

52 Teilhard de Chardin, *Phenomenon of Man*, 140.
53 Peter B. Medawar, "VI.—Critical Notice. *The Phenomenon of Man*. By Pierre Teilhard de Chardin. With an introduction by Sir Julian Huxley. Collins, London, 1959. 25s," *Mind* 70, no. 277 (1961): 99.
54 Richard Dawkins, *Unweaving the Rainbow: Science, Delusion and the Appetite for Wonder* (Boston: Houghton Mifflin, 2000), 185.
55 Julian Huxley, "Introduction," in Teilhard de Chardin, *Phenomenon of Man*, 26.

Except in the Light of Evolution" (1973) echoes Teilhard de Chardin's conviction that there is a self-directedness in nature, was a great admirer too, naming him "one of the great thinkers of our age."[56]

But if Teilhard de Chardin's premise that the petrified intelligence of the earth's rocky strata possessed a latent spiritual force was seen as flaky by scientists of the mechanistic worldview, it was simultaneously deemed so nonconformist by his Christian contemporaries as to outright suppress the distribution and dissemination of his theological musings. At the other end of the spectrum, then, the Catholic Church initially also took a dim view of Teilhard de Chardin's work, an attitude that would only begin to change after his death. Indeed, during the course of his lifetime, the Vatican took measures to prevent the influence of his efforts to bring together evolutionary science with Christian faith. In 1926, his Jesuit superiors prohibited him from teaching, and even though the Vatican did not go so far as to declare *The Phenomenon of Man* heretical, it was nevertheless barred from publication when he submitted it to Rome for approval. Seven years after his death, the Vatican issued a monitum about the success of his posthumous publications, warning seminary rectors and university presidents to protect the minds of the youth against its offense to Catholic doctrine. Challenging a literal reading of the book of Genesis, Teilhard de Chardin's unorthodox interpretation of original sin eventually proved too much for some Church officials to accept. But the Vatican's censorship was presumably motivated too, and no less fundamentally so, by the skepticism among theologians about evolutionary theory and the concomitant world-affirming perspective that Teilhard de Chardin, as a result, brought into his understanding of Christianity.[57] Yet, in spite of the measures taken, *The Phenomenon of*

56 Theodosius Dobzhansky, "Nothing in Biology Makes Sense Except in the Light of Evolution," *American Biology Teacher* 35, no. 3 (1973): 129.

57 Linda S. Wood, *A More Perfect Union: Holistic Worldviews and the Transformation of American Culture after World War II* (Oxford: Oxford University Press, 2010), 126.

Man rapidly became a bestseller, and at least some of the officials and experts had most likely been exposed to Teilhard de Chardin's "dangerous ideas," for by 1965, less than three years after the monitum, his Christian interpretation of modern scientific concepts had become widely known and even appreciated, to such a degree that even the Church had officially come to adopt some of the startling claims as its own.

The publication of Vatican II's *Pastoral Constitution on the Church in the Modern World (Gaudium et spes)* (1965) served as a turning point in this case. In particular, it gave affirmation to a dynamic and evolutionary conception of reality in place of a static world picture, but it also acknowledged that the modern turn from stasis to dynamism called for the analysis and synthesis by Catholics of a number of new problems, including its emphasis on the importance of intervening duties in the phenomenal world, which, within an evolutionary framework, rather than being diminished by a hope related to the eschaton, undergirds the importance of such worldly efforts.[58] It is impossible not to compare the content of this document with some of the key ideas of Teilhard de Chardin's writings, such as his basic claim that the worldly domain of creation is becoming rather than being, and that it is therefore subject to an organic future of persistent self-production. From a Teilhardian point of view, one would be amiss not to make out a belated response by the Church to the modern accusation that the Christian virtue of hope constitutes but an escapism from worldly affairs, and, worse, a fatalism as to the predetermined outcome of our every action. But once we heed Teilhard de Chardin's evolutionary understanding and realize that creation is a work-in-progress, authentic hope will vehemently orient us toward a poietic participation in the divine plan that is working to actively produce the future. We might rightly ask, however, What it is that makes humankind so unique according to Teilhard de Chardin? And what it is that leads him to claim that this insight of uniqueness

58 Vatican II, *The Pastoral Constitution on the Church in the Modern World* (Gaudium et spes) (Rome: The Vatican, 1965), 5.

coincides with the withering away of the Cartesian subject? In order to answer these questions, we must delve deeper into his organicist ontology, and in particular the central role ascribed to technology within it.

Science and Seeing

To see and to make others see — such is the objective of *The Phenomenon of Man,* if we are to believe Teilhard de Chardin himself. "The whole of life lies in that verb," he writes about the active form of perception.[59] What is it that he wishes for us to perceive, though? Put simply, it is "the unfolding of the whole," and the whole that Teilhard de Chardin writes of is nature, which is ontologically constituted by an evolutionary process of becoming. But catching sight of this cosmogonic insight, according to Teilhard de Chardin, entails that we do not merely fix our gaze on nature's myriad surfaces — the domain of matter — but turn our eyes toward its unified interior too, which is where the domain of mind comes into play. Still, what kind of seeing reveals, aside from the surfaces of things, also what lies within? According to Teilhard de Chardin, intertwined with this question is a further one: Does nature operate like an automaton, without a purpose of its own, or is it rather self-directed, with "an absolute direction of growth"?[60] Although there at first glance seems to be no obvious connection, these two questions — how the within can be perceived, on the one hand, and whether evolution has an aim, on the other — are closely connected. As he puts it, we cannot comprehend the latter until we have gazed into the abyss of the former, and it is only when participating in the latter that the former becomes perceivable. We must attune our senses in such a manner that the appearance of nothing but the blind mechanism of efficient cause reveals a directed harmony concealed beneath. Indeed, efficient cause

59 Teilhard de Chardin, *Phenomenon of Man,* 31.
60 Pierre Teilhard de Chardin, *Writings in Time of War* (New York: HarperCollins, 1968), 32.

cannot possibly be exhaustive, because, as Teilhard de Chardin argues, "the world does not hold together 'from below' but from 'above.'"[61] As is often the case in Teilhard de Chardin's writings, this is a statement with both theological and philosophical undertones. However, connected to the notion of seeing, what he means by the word is considerably more extensive than mere visual sight. Mirroring the Romantics, Teilhard de Chardin argues that coexistent with the "without" of nature, there is also a "within," and the process of evolution, in its universal sense, is one of increasing articulation of self-consciousness: it does not solely consist of novel and complex forms of the organization of matter, but equally of novel and complex forms of the organization of mind. Evolution cannot simply be conceived of in terms of a linear progression of increasing complexity in the organization of matter, but consists, rather, "in a continual tension of organic doubling-back upon itself, and thus of interiorization."[62] It is, in other words, as much an evolution of the within as it is an evolution of the without.

Consequently, Teilhard de Chardin believes that the most fundamental and profound manner of comprehending the history of the earth is to describe the unfolding of the spirit of the earth,[63] that is, the earth coming to self-consciousness of itself, which he defines as an internal tendency toward "the affinity of being with being."[64] Although his is a mystic's depiction of natural evolution, in the sense that it is largely a retelling of the Christian narrative of the restoration of spirit from its fallen state back into union with God, the symbolism can arguably be cashed out in naturalistic terms, if by naturalism we mean the rejection of the existence of beings outside of the domain of nature. Because what Teilhard de Chardin primarily takes a stand against is mechanistic philosophy, which concerns itself solely with motion, or the collision of objects and the exchange

61 Pierre Teilhard de Chardin, *Christianity and Evolution* (New York: Harvest, 1974), 113.

62 Teilhard de Chardin, *Phenomenon of Man*, 301–2.

63 Ibid., 245.

64 Ibid., 264.

of force. The methodological implications of viewing the earth as an automaton, he laments, is that that explanation is thought to have everything to do with the determinism of inert matter and nothing whatsoever to do with the spontaneity and free will of the living spirit. In modern scientific terms, the latter is only associated — if its existence is acknowledged at all — with human consciousness, deemed an odd exception, an aberrant function, or, as is most often the case, an epiphenomenon. But as Teilhard de Chardin points out, there can be no systematic explanation of the unity of nature as long as spirit is exclusively taken to be but "an erratic object in a disjointed world," since "man, in nature, is a genuine fact falling (at least partially) within the scope of the requirements and methods of science."[65] In opposition to the Kantian critique of metaphysics, which, by proceeding from the Cartesian separation between *res extensa* and *res cogitans,* had only further cemented modern science as exclusively the measurement of physical phenomena, Teilhard de Chardin's endeavor to situate spirit within the same domain that modern science seeks to render comprehensible to reason directs our attention to the technical artifacts and practices that mediate sensory experience in the first place.

Although Teilhard de Chardin, in spite of Kant's prohibition on human access to the thing-in-itself, provocatively calls his own effort a "scientific" project,[66] he does so in order to broaden the meaning of the term to encompass a phenomenology that is dissatisfied with reducing science to the kind of empiricism that proceeds from objects as they appear to consciousness. His intention is to "break through and go beyond appearances"[67] to the intuitive source of our seeing. As such, the project in question is not at all unlike the concern of second-order cybernetics to establish a universal science of observing systems, that is, to observe how the scientific gaze itself operates, a concern that,

65 Ibid., 34.
66 Ibid., 29.
67 Pierre Teilhard de Chardin, *Letters from a Traveler* (New York: Harper & Row, 1962), 70.

as we shall see in chapter 4, would properly come to the fore in the theory of knowledge during the second half of the twentieth century. In other words, the modern trend against which Teilhard de Chardin reacted was one of greater isolation and alienation from nature caused by humanity's further retreat into solitary confinement, which, for him, was typified by the technical image of a machine or an engineer who passively transforms sensory input into a finished product, but who does so always with an implied ontological distance between creator and created. The theologian Thomas King writes that "in placing man [in the framework of phenomenon and appearance], Teilhard does not mean the flat veneer of colors that strike the retinas. Rather he wants to show the meaning that haloes man when he is placed in the context of a vast cosmic movement."[68] "Perception," as he uses the term, is thus not to be confused with the passive registering of sensory impressions by an aloof subject, but has rather to do with an active bringing-forth in the world — what he calls a "cosmic sense" — that he locates in nature's "internal propensity to unite."[69] Although he bends over backward, in *The Phenomenon of Man,* to assure the reader that what he lays out therein is neither a theory of knowledge nor a theory of being, it is possible to make the case that Teilhard de Chardin's is an attempt to demonstrate how the instrumental rational attitude presumes a more fundamental, poietic relation to the world, which comes to expression in the retrieval of the organic and its concomitant development of the intrinsic-teleological notion of immanent self-organization.

Although Teilhard de Chardin championed the findings of the modern scientific disciplines of astronomy, geology, and biology post-Kant, he did not accept, at face value, the modern attitude that had walled off conscious spirit from the machine-like determinism of mechanistic nature. Quite to the contrary,

68 Thomas M. King, *Teilhard's Mysticism of Knowing* (New York: Seabury Press, 1981), 46.

69 Teilhard de Chardin, *Phenomenon of Man,* 264.

he saw in the history of the earth the procedure of a spiritual unfolding of absolute significance for the material world:

> Until the dawn of the present era, one could say that man still had the illusion of living "in the open air" in a universe that was penetrable and transparent. At that time there was no hard and fast boundary, and all sorts of exchanges were possible between the here below and the beyond, between heaven and earth, between relative and absolute[. ...] Then, with the rise of science, we saw the gradual spreading over everything of a sort of membrane that our knowledge could not penetrate.[70]

By referring to "the dawn of the present era," Teilhard de Chardin in effect equates the birth of modernity with Kant's transcendental idealism, such that the impenetrable "membrane," in this context, can be understood as the transcendental restrictions with which Kant had limited the heretofore free reign of pure reason. "Thus I had to deny knowledge," Kant famously declared, "in order to make room for faith."[71] In the wake of Kant's self-proclaimed Copernican revolution, the purposefulness of spirit in nature had been relegated from being constitutive of the geocosm into maxims of human thought and action. In other words, it was no longer possible for humankind to intervene into the astral ecology that ties together heaven and earth. The natural scientific method, in its mechanistic brand, had not troubled itself with "the within" of nature — either outright denying its relevance for humanity's understanding of nature or rejecting it as an epiphenomenal illusion — but Teilhard de Chardin, in contrast, saw "the within" and "the without" — or, put differently, "spirit" and "nature" — as so interdependent that he wondered if scientists ever managed to objectively expose a so-called external and objectified world entirely apart from

70 Pierre Teilhard de Chardin, *Activation of Energy: Enlightening Reflections on Spiritual Energy* (New York: Harper Collins, 1978), 186.

71 Kant, *Critique of Pure Reason,* 117 (Bxxx).

themselves, "or quite simply and unconsciously [...] recognized and expressed themselves in it."[72] Even though he praised the Baconian method of modern science, declaring the *Novum Organum* a "marvelous instrument [...] to which we owe all our advances,"[73] Teilhard de Chardin nevertheless argued that, by breaking the human down to nothing but a pile of dismantled machinery—what might at first glance look like a rather humble and nonanthropocentric perspective on our condition—it had in effect glossed over how this supposedly objective panoptic still remained instrumentalist on the most basic level of its attitude toward nature.[74] His point is that radical materialists too, even though they are right to incorporate mind as a phenomenon of the world rather than being fundamentally exterior to it, have a tendency to simplify this relationship. Whenever cognition and reason are reduced to the epiphenomenal illusion of underlying mechanism, we tend to solve only half of the equation and forget the rest, namely, how reason participates in the same evolutionary process by actively molding the world in understanding.

What modern science reveals, according to Teilhard de Chardin, is that we are not beings isolated from nature by way of certain transcendental ideal limits. It is rather through our participation in the geological economy of the earth that we—the object and subject conjoined—become what we essentially are. These findings, he concurs, should lead us to question our previous understanding of ourselves. But it does not follow from this that we should therefore also question our own significance. The challenge it poses to the modern separation between the spiritual and the material is not something to be lamented, but something to be celebrated—by science and theology alike. Being *a part of* rather than *apart from* the earth is to be implicated in the unfolding of a greater whole. No longer will science

72 Pierre Teilhard de Chardin, *The Vision of the Past,* trans. John M. Cohen (New York: Harper & Row, 1966), 69.

73 Teilhard de Chardin, *Phenomenon of Man,* 258.

74 Ibid.

be able to pretend not to deal with questions of meaning, just like theology will no longer get away with turning its back on worldly affairs.[75] If religion still has a role to play in our modern society, it is for Teilhard de Chardin through the championing of worldly concerns. And conversely, if the progress of science is to avoid a reactionary and fundamentalist response to the nihilistic consequences of mechanism, it must be because its own capacity to provide us with a meaningful existence renders such a fundamentalism superfluous. We must commit to disciplines such as geology not only the unearthing of the natural history of our planet, but also the unearthing of ourselves as part of that history.

It is important to note that Teilhard de Chardin did not reject the centrality of sensory experience for the scientific method. On the contrary, he stresses the importance of the without, because it is only when the human proceeds "out of himself [and] into the immensity and dangers of the universe" that he can intuitively feel the awakening of the cosmic sense within, which, according to Teilhard, consists in the realization that "object and subject marry and mutually transform each other in the act of knowledge."[76] Proceeding from the assumption that knowledge implies an objective model of reality, René Descartes had abstracted the scientific enterprise out of the natural world that constitutes it. But such a representationalist notion of objectivity fails to account for the ways in which the human is caught body and soul in the technical practices through which it encounters nature in the first place. In *Activation of Energy* (1963), a collection of essays written between 1939 and 1955, Teilhard de Chardin takes the decentering that humans suffered in modernity—first from the center of the universe in the wake of Copernicus, then from the center of the living world in the wake of Darwin, and last but not least even from the innermost core of their own selves in the wake of Freud—as proof of his claim that not only is the object actively shaped in the work of

75 Ibid., 56.
76 Ibid., 32.

the subject, but the subject, too, is actively molded and worked upon in its confrontation with the object: "The most agonizing experience of modern man, when he has the courage or the time to look around himself at the world of his discoveries, is that it is insinuating itself, through the countless tentacles of its determinisms and inherited properties, into the very core of what each one had become accustomed to calling by the familiar name of his soul."[77] No longer self-assuredly standing at the center of an unchanging natural order, humans were forced to reassess their being in the wake of the crisis of the meaning-making mythologies of the premodern Occident. But contrary to the conventional historiography that finds in the Copernican, Darwinian, and Freudian discoveries a disenchantment with the world in the form of three disorienting acts of decentering, Teilhard de Chardin sees these figures as heralds of a modern development leading to the rediscovery of humankind's import, not through a conception of humanity championed by Enlightenment humanism, but through a reconceptualization of what it means to be human in the wake of modernity. There are no reasons to believe that we cannot complement the Old Testament's praise of creation with the findings of modern science because, according to Teilhard de Chardin, its wonders were even more vividly revealed in the wake of Copernicus. Consequently, Teilhard de Chardin pleads with his readers to look again at what modern science has shown us, and to look at it *anew:*

> With the mere admission of a revolution of the earth around the sun; simply, that is by introducing a dissociation between a geometric and psychic center to things — the whole magic of the celestial spheres fade away, leaving man confronted with a plastic mass to be re-thought in its entirety. It was like the caterpillar whose substance (apart from a few rare cerebral elements) dissolves, as its metamorphosis draws near,

77 Teilhard de Chardin, *Activation of Energy*, 187.

> into a more or less amorphous product: the revised protoplasmic stuff from which the butterfly will emerge.[78]

Dissolving the antique division between the divine celestial realm, on the one hand, and the fallen terrestrial realm, on the other, Teilhard de Chardin celebrates the Copernican revolution for the way in which it set humans upon a path toward the completion of the historical process of spiritual incarnation in the terrestrial realm.

Anticipating the speculative wager of twentieth-century cosmology — that ours is a topologically flat and infinitely vast universe, "whose center is everywhere and whose circumference is nowhere"[79] — Teilhard de Chardin describes how "man is seen not as a static center of the world — as he for long believed himself to be — but as the axis and leading shoot of evolution, which is something much finer,"[80] since it means that we as humans do not merely receive sensory impressions of a finished product, but actively participate in the becoming of a single organism that is evolving toward increased complexity in accordance with its final cause. Although humanity contributes by conferring upon the universe a "form of unity it would otherwise (i.e. without being thought) be without,"[81] it is not simply a matter of human cognition producing artificial representations of some real state of affairs in the outside world of objective nature, but the immanent unfolding of self-consciousness as an incessant movement of becoming that constantly pushes the self beyond its own limits "to become part of a growing common movement of life."[82] Applying the tools of reason, we are not fated to produce a mere representation of the earth in our own image. By

78 Ibid., 254.

79 Incidentally, this is also how some of the more esoterically inclined scholastics is said to have characterized God. See Jorge L. Borges, "The Fearful Sphere of Pascal," in *Labyrinths: Selected Stories & Other Writings* (New York: New Directions, 1962), 189–92.

80 Teilhard de Chardin, *Phenomenon of Man*, 36.

81 Ibid., 249.

82 King, *Teilhard's Mysticism of Knowing*, 26.

cultivating a poietic attitude of receptiveness to nature's disclosure, we may ourselves become the conduit through which the earth comes to know itself. Teilhard de Chardin describes the earth's striving for self-consciousness and the concomitant perfection of spirit in nature as "the definitive access of consciousness to a *scale of new dimensions;* and in consequence the birth of an entirely renewed universe, without any change of line or feature by the simple transformation of its intimate substance,"[83] thus he likens the working of reason to nature's organization of itself into ever-more complex forms of life — be they natural or artificial. Nature and spirit are mirror images of each other that both diverge out of and merge into a joint source, beginning in an unconscious harmony, as spirit is fully implicit nature, and ending in the conscious revelation of the same harmony once again, after nature has been made explicitly artificial. History is situated within either of these ends, as a negotiation between subject and object in an effort to understand how both "hold together and are complementary."[84] Contrary to the Cartesian substance dualism of mechanistic philosophy, Teilhard de Chardin thus views humanity as part of nature's self-organization too. Hence, the Copernican revolution was not the crippling humiliation Freud wanted it to be.[85] It could instead be made — as in the work of Teilhard de Chardin — to account for an extension of vitality throughout the entire structure of the universe.

Machinic Lamarckism, or, a Political Theology of Transformism

As we have seen, what made Teilhard de Chardin's concept of the univocity of being unique was his evolutionary perspective. It was during his stay at Hastings, England, in the first decade of the twentieth century, as he was completing his theological

83 Teilhard de Chardin, *Phenomenon of Man,* 219.

84 Ibid., 63.

85 Thomas Moynihan, *X-Risk: How Humanity Discovered Its Own Extinction* (Falmouth: Urbanomic, 2020), 83–84.

studies, that Teilhard de Chardin first read Bergson's *Creative Evolution* (1907). Although Darwin's *On the Origin of Species* (1859) and *The Descent of Man* (1871) proved important for his lifelong commitment to an evolutionary outlook, it was rather Bergson's proposal of an orthogenic alternative to the mechanism underlying Darwin's theory of natural selection that convinced Teilhard de Chardin that species, including our own, are mutable throughout natural history, and most importantly, that this evolutionary process ought to be understood as more than mere mechanism without telos. As pointed out by Monod, the idea "according to which the evolution of the biosphere culminating in man would be part of the smooth onward flow of cosmic evolution itself [...] did not originate with Teilhard." On the contrary, "it is in fact the central theme of nineteenth-century scientific progressism" — a "single principle," thanks to which "man at last finds his eminent and necessary place in the universe, along with certainty of the progress which is forever pledged to him."[86] Indeed, orthogenesis can be understood as an Aristotelean rejoinder to the unfettered march of mechanistic philosophy into the domain of the life sciences, rejecting blind mechanism as the exclusive determinant of evolution, but without necessarily restricting the directedness to the idea of a preformed individual.

Until at least the mid-nineteenth century, as demonstrated by the historian of philosophy Étienne Gilson, the term "evolution" exclusively denoted what would later come to be known as "ontogenesis," that is, the development of an organism from the point of fertilization onward, akin to the actualization of an individual from out of the potency latent in its very origin. It sought to address the question of how to allow for the explanation of the gradual development of organisms without simultaneously infracting upon the omnipotence of the divine artificer as its first cause. Although certainly anticipated already in the work of Descartes, Comte de Button, and Kant, the theory of evolution familiar to twenty-first-century biology was

86 Monod, *Chance and Necessity*, 32–33.

widely received first after Lamarck and Darwin had ventured the hypotheses that organisms too are inextricably involved in a vast process of becoming, in which every being, rather than constituting fixed points in a hierarchic pecking order, is seen to emerge out of previously existing forms in a network of indivisible continuity, and, moreover, that such transformations were not solely the result of an internal generation in the individual organism, but were also the product of external, environmental effects. Prior to this paradigmatic shift, the early doctrine of evolution held that the development of the organism was but the "un-rolling of something already given,"[87] and hence the idea that species themselves may undergo change was not considered. In this sense, the development or transformation of an organism referred to the visibility and palpability of what had previously been invisible and impalpable. Everything that ever had been, or would be, already existed in its latent form, such that it was merely a question of its actualization.[88] Posing the question of the origin and transformation of species, however, Lamarck and Darwin came to challenge the fixed nature of the individual organism, and in effect to alter the very meaning of evolution that had dominated since the scholastic adoption of the Aristotelean formal cause. In other words, their idea of a transmutation of species — or simply "transformism," as it is sometimes known — refused to accept as given that species are characterized by an eternal constancy in their nature. In the summary of chapter 6 in the second part of his *Zoological Philosophy* (1809), Lamarck wrote that

> since all living bodies are productions of nature, she must herself have organized the simplest of such bodies, endowed them directly with life, and with the faculties peculiar to living bodies. [And] by means of these direct generations

87 Étienne Gilson, *From Aristotle to Darwin and Back Again: A Journey in Final Causality, Species, and Evolution*, trans. John Lyon (Notre Dame: University of Notre Dame Press, 1984), 50.
88 Ibid., 49–51.

> formed at the beginning both of the animal and vegetable scales, nature has ultimately conferred existence on all other living bodies in turn.[89]

Although they change more slowly than humans can directly observe, species are not subject to their fixed nature but constantly changing and becoming different, a transformation that Lamarck sought to explain in terms of variations in the surrounding environment.[90]

In this particular regard, Lamarck's and Darwin's respective theories were largely in agreement with each other. But whereas Darwin's theory of natural selection appealed solely to a series of random hereditary variations resulting — blindly and purely by chance — in the best adaptation of a population at a given time and in a given environment, Lamarck on the contrary ascribed to organisms the capacity to actively modify themselves in order to adapt to changes in their surroundings, and in so doing establish a relationship of interaction with their environment in terms of a dynamism between the development over time of their needs and their habits:

> It is not the shape either of the body or its parts which gives rise to the habits of animals and their mode of life; but that it is, on the contrary, the habits, mode of life and all the other influences of the environment which have in course of time built up the shape of the body and of the parts of animals. With new shapes, new faculties have been acquired, and little by little nature has succeeded in fashioning animals such as we actually see them.[91]

As opposed to Darwin, who attributed no autonomy to the organism itself, Lamarck held that an organism adapts itself to

---

89 Jean-Baptiste de Lamarck, *Zoological Philosophy, or Exposition with Regard to the Natural History of Animals*, trans. Hugh Elliot (New York: Hafner Publishing, 1963), x.

90 Ibid., 107–12; Gilson, *From Aristotle to Darwin and Back Again*, 44.

91 Lamarck, *Zoological Philosophy*, 127.

external conditions by making "more frequent use of some of its parts which it previously used less, and thus greatly [developing] and [enlarging] them; or else to make use of entirely new parts, to which the needs have imperceptibly given birth by efforts of its inner feeling."[92] Although this notion is heavily indebted to the Aristotelian conception that living beings, "working from within by their substantial form, progressively shape their matter according to the type of perfected being which they tend to become,"[93] Lamarck found the immanent tendency toward perfection in the generation of complex patterns of symbiosis in nature at large. The Lamarckian view is one that, although it dispenses with the idea that each species develops out of a ready-made blueprint provided by a divine creator, still manages to maintain the purposive directedness of the divine spirit by immanentizing it within the interior abyss of the earth's infinite productivity, such that, "even if one situates it initially in the mind of God, it would be rather necessary to conclude by rediscovering it [in nature],"[94] or by demonstrating how mind and matter coincide in nature's capacity to self-organize.

We see, here, the Lamarckian influence upon Teilhard de Chardin's understanding of an internal propensity in nature to purposively organize itself. One might say that evolution, for Teilhard de Chardin, is a fundamental aspect of reality—an ontological condition to which modern science must adapt if it is henceforward to explain the world in any satisfactory sense.[95] Indeed, "Teilhard's ascending energy," Monod has argued,

> is a plain instance of animist projection. In order to give meaning to nature, so that man need not be separated from it by a fathomless gulf, and for it again to become decipherable and intelligible, *a purpose had to be restored to it.* Should no spirit be available to harbor this purpose, then one inserts

92 Ibid., 112.
93 Gilson, *From Aristotle to Darwin and Back Again*, 46–47.
94 Ibid., 48–49.
95 Ibid., 219.

> into nature an evolutive, an ascending "force," which in effect amounts to abandoning the postulate of objectivity.[96]

But Teilhard de Chardin is keen to argue that his is not merely a metaphysical presupposition, for he is at a loss to understand, even from an empirical point of view, how evolution could be conceptualized without accepting the presence of some inherent progressive tendency. Whichever geosphere we consider—whether it be the rocky layers that make up the earth's crust, the diversity of lifeforms that dwell upon its surface, or the layer of gases surrounding it—the conclusion is unmistakable: all physical processes are subject to the deep time of geological history, and the earth as a systematic whole is incomprehensible except through such an evolutionary lens.[97] Zoological evolution, then, is merely one instance of natural evolution in general. Moreover, anticipating the French philosopher Georges Canguilhem's critique of mechanism, Teilhard de Chardin believes that such an ontology can coherently portray nature as the mindless motion of external force only by unconsciously repressing the role played by mind in explaining it. On the contrary, not even mind can, in Teilhard de Chardin's eyes, be beyond the becoming of the world. Then again, he does not reject the evolutionary mechanism posited by Darwin, but merely finds that mechanism alone is insufficient to explain what we empirically observe:

> In various quarters I shall be accused of showing too Lamarckian a bent in the explanations which follow, of giving an exaggerated influence to the *within* in the organic arrangement of bodies. But be pleased to remember that, in the "morphogenetic" action of instinct as here understood, an essential part is left to the Darwinian play of external forces and to chance. It is only really through strokes of chance that life proceeds, but strokes of chance which are recognized

96 Monod, *Chance and Necessity,* 33.
97 Teilhard de Chardin, *Vision of the Past,* 245–47.

> and grasped—that is to say, psychically selected. Properly understood the "anti-chance" of the Neo-Lamarckian is not the mere negation of Darwinian chance. On the contrary it appears as its utilization. There is a functional complementariness between the two factors; we could call it "symbiosis."[98]

Consequently, Teilhard de Chardin's understanding of evolution is explicitly teleological. Natural evolution is like a line of reasoning gradually winding its path forward, over time becoming increasingly self-directed and intentional, such that even if humanity cannot foresee where evolution will take it, this is not the same as saying that its evolution is completely random.[99] It is as much a "holding together" from over and above as it is a "stumbling around" from under and below. Nature unfolds in accordance with its own organization of itself, and part of its unfolding is an increased directedness in the manner by which it unfolds.

§

Considering his metaphysics, then, it is not at all surprising, but still informative to point out, that Teilhard de Chardin imagines technological evolution as merely yet another instance of natural evolution. Technology is not only conceived of as thoroughly natural, but as a crucial component for the teleological unfolding of nature. "Is it not precisely the world itself," he rhetorically asks, "which, culminating in thought, expects us to think out again the instinctive impulses of nature so as to perfect them?"[100] Any dichotomous opposition between biology and technology is thus essentially untenable, for although the machine emerges first with the artful skill of modern humans, the artificial organ is still, on an ontological level, functionally equivalent with the natural organ. Indeed, organisms are already dependent, in their

98 Teilhard de Chardin, *Phenomenon of Man*, 149.
99 Teilhard de Chardin, *Vision of the Past*, 181.
100 Teilhard de Chardin, *Phenomenon of Man*, 283.

physiology, upon organs internal to their corporeal selves: natural organs — limbs, hearts, kidneys, and so on — whose function, should they cease to work as intended, may be replaced with artificial organs — prostheses, ventricular assist devices, haemodialysis machines — or even enhanced by artificially improving upon their particular mechanism. The Deleuzo-Guattarian concept of the body as a "machinic assemblage"[101] is particularly suitable as a description of what Teilhard de Chardin has in mind, not the least since it brings to our attention the notion of the invariably scalar reflection of the whole (the macrocosm) in the part (the microcosm), and vice versa.[102] In this sense, the artificial world of human beings is a microcosm whose composition and structure correspond to that of the environment in which it is situated, namely, the earth and, in turn, the universe. Like the body of the biological organism, Teilhard de Chardin imagines that the body of the geophysical organism is dependent upon its own organs. What is more, because we are faced with this scalar invariance, the ontological question of the essential boundary between organism and environment is reduced to the pragmatic question of analytical convenience. This is the manner in which we may understand Teilhard de Chardin's contention that although artificial organs extend or project the human microcosm upon the rest of his terrestrial environment, this is not the same as to argue that it therefore also instrumentally enframes the earth, an outside world crafted in the image of the human, as Kant would have it. For insofar as there is a deeper and more fundamental harmony between nature and spirit, the aforementioned projection of artificial organs cannot be said to be any more or less instrumental than natural organs. Initially, following the spontaneous formation of patterns through processes of self-organization in nature, biological reproduction becomes the most efficient means by

---

101 Gilles Deleuze and Félix Guattari, *A Thousand Plateaus: Capitalism and Schizophrenia*, trans. Brian Massumi (Minneapolis: University of Minnesota, 1987), 34–38, 71–74, 88–90.

102 Thomas M. King, *Teilhard de Chardin* (Wilmington: Glazier, 1988), 83.

which complexity is proliferated. But along with the advent of humankind's technological manipulation of nature, we are faced with yet another paradigmatic shift to the self-directedness and acceleration of Lamarckian evolution, as changes in the form of acquired characteristics can now be transmitted actively, by artificial means, rather than only passively, through genetic information. From this point onward, evolution operates not primarily as a natural-biological phenomenon but rather as an ideational-artifical phenomenon. In modernity, spirit is able to deliberately reorganize nature, and to do so on a planetary scale.

To illustrate the planetary effect of spirit, which had emerged quite imperceptibly over the relatively short period of human history, at least compared to the macro-scale of deep time established by Hutton, Vernadsky noted that "the mineralogical rarity, native iron, is now being produced by the billions of tons[. ...] The same is true with regard to the countless number of artificial chemical combinations (biogenic 'cultural' minerals) newly created on our planet."[103] In depicting the importance of this artificial system of transmutation, Vernadsky did not simply seek to complement the geological economy with just another variable. We would be mistaken in regarding his concern for humanity's technical effect on the earth as forming no more than an added sum, because along with the technical means by which substances can be artificially altered and created, Vernadsky finds that the earth itself is qualitatively transformed — it introduces a structural shift to the functional manner in which the geospheres interact. We are therefore faced with the emergence of an entirely novel sphere enveloping the earth, what Vernadsky and Teilhard de Chardin jointly call "the noösphere," attaching as a prefix the Greek word for "mind" or "spirit." Just like 3.5 billion years ago, when a thin biotic film began to cover the surface of the lithosphere, so modernity, Teilhard de Chardin argues, marks another decisive stage in the history of the earth. Via the global proliferation of technology, yet another concentric geosphere is added, which, besides the

103 Vernadsky, "The Biosphere and the Noösphere," 9.

obvious noetic processes (calculating, modeling, communicating, deliberating) involves physical infrastructures through which these processes are carried out, such as artifacts, devices, organizations, and techniques. This stage in natural evolution signals something like a distributed intelligence, the technological materialization of spirit, conceived as an extended, externalized, and institutionalized structure.

Humankind is thus the site of the evolutionary shift from natural selection to invention. Humans are but one family of primates among others in the animal kingdom, Teilhard de Chardin admits, but yet, by giving rise to the noösphere, humankind nevertheless also marks a "phase transition"[104] — a metamorphosis of the earth — because of its ability to decipher, transform, and rewrite the essence of life. Spirit, with the emergence of the noösphere, is now able to perceive its own logic in the noumenal essence of organic nature. This we see, according to Teilhard de Chardin, in the prodigy of physicists manipulating nature on the atomic scale, the manner in which geneticists are making us envisage the possibility of editing chromosomes, or in our medical ability to modify the powers of the organism

104 Although "phase transition" is used in a more delimited and precise sense in physics to describe transitions between solid, liquid, and gaseous states of matter, I have borrowed a more inclusive sense of the term from the assemblage theory of the philosopher and architectural theorist Manuel Delanda, wherein it denotes the more general manner in which intensive genesis creates forms through difference, spontaneously activating *qualitative* change when critical differences between indivisible, intensive processes of quantitative character such as temperature, speed, and pressure occur. See Manuel Delanda, *Assemblage Theory* (Edinburgh: Edinburgh University Press, 2016), 19. By referring to nonquantifiable, productive differences in which forms become other forms at critical points of intensity, such as when water turns into ice, or when cell differentiation produces organisms, an ontology of intensive processes thus thinks of difference not as the failure of a preformed object to be another, but as an intrinsic part of any object. This is also the manner in which I propose that we understand Teilhard de Chardin's shift in natural evolution related to the emergence of the noösphere, namely, as a phase transition into a different qualitative state resulting from an increased intensity in what Teilhard calls nature's "radial energy." See Donald P. Gray, "Teilhard's Energy," *CrossCurrents* 21, no. 2 (1971): 238–40.

through hormones.[105] In laboratories, life is becoming reproducible. Molecular characters are entering into a new, technological milieu, such that passive heredity assumes a noöspheric form.[106] Physical processes are transformed into concepts, and in vivo biomolecules transmute into in silico symbols, so that heredity itself becomes hominized.[107] Whereas animals adapt to environmental challenges via learning, humans, Teilhard de Chardin argues, not only learn but know *that* they learn, and *how* they learn, and may thus improve their capacity for learning.[108] Humankind represents the latest transition from an unconscious participation in evolution via a self-conscious, manipulative understanding of these mechanisms, putting them to work on behalf of conscious planning, even assuming responsibility for the future course of evolution as such, thereby sublating the boundaries between the natural and the artificial by giving rise to synthetic hybridization. Indeed, in modernity, evolution and selection are being transposed from the biosphere into the noösphere. Passive and slow, natural evolution is sublated into a conscious, accelerated, and progressively systematic global endeavor. Artifice gradually takes upon itself the task of the natural, insofar as the transmission techniques of a literate culture — techniques for reading, editing, and rewriting archives of symbolic materials — are superimposed onto genetic heredity. The forms of self-organization we call organisms have thus brought forth into nature the capacity to consciously alter themselves, such that a Teilhardian interpretation of the meaning of the Anthropocene would be to propose that it marks our awakening to the idea of a proactive and synthetic idea of evolution. With the planet so constructed, humankind, in its artistic role as the conduit of artifice, is in the aporetic position of being simultaneously determined by, and responsible for, the evolutionary progressivism behind modernity's global technological systems.

105 Teilhard de Chardin, *Activation of Energy,* 160.
106 Teilhard de Chardin, *Phenomenon of Man,* 225–26.
107 Ibid., 247.
108 Ibid., 168.

Humans, once they begin to grasp the underlying laws of all systems, natural and artificial, find their expression of freedom, in their purposive construction of technology, to be perfectly in line with the unconditioned productivity of nature at large.

Indeed, Vernadsky too strongly believed in evolution as an inevitable and teleologically inclined tendency toward an increased integration of humankind's artificial products with the earth's natural products, such that any negative side effects caused by the expansion of the noösphere would eventually be justified by a kind of theodical promise of salvation through technological transformation. Through artifice's transition from the stage of anticipation toward an actual, comprehensive life-creation, achieving a quality of full organicism by mastering the transformation of energy by living matter, humanity as a whole would be able to manipulate and reproduce the planetary condition by means entirely different from those of unconscious nature.[109] In this sense, the noösphere would constitute but "the last of many stages in the evolution of the biosphere in geological history."[110] Indeed, Vernadsky would take this conclusion so far as to argue that the progressive geological evolution (on a time scale of billions of years) and the war against regressive, fascist barbarism (during the twentieth century) were fundamentally related:

> Now we live in the period of a new geological evolutionary change in the biosphere. We are entering the noösphere. This new elemental geological process is taking place at a stormy time, in the epoch of a destructive world war. But the important fact is that our democratic ideals are *in tune with* the elemental geological processes, with *the laws of nature,* and with the noösphere. Therefore we may face the future with confidence. It is in our hands. We will not let it go.[111]

---

109 Anastasia Gacheva, "Art as the Overcoming of Death: From Nikolai Fedorov to the Cosmists of the 1920s," *e-flux* 89 (2018): 10–11.
110 Vernadsky, "The Biosphere and the Noösphere," 9.
111 Ibid., 10 (my emphasis).

Importantly, though, from Vernadsky's perspective, the progressivism in question cannot really be said to be the result of humankind's ability to instrumentally take control over nature. On the contrary, nature is expressing itself *in* artificial products, realizing itself *through* technical means. In fact, this is the reason for why he may argue that there is no use in resisting the emergence of the noösphere, for the technical organization of ever-greater forms of complexity is an expression inherent to nature as such. "In a historical contest, as for instance in a war of such magnitude as the present one," Vernadsky declares with great confidence, "he finally wins who follows that law [that is, 'the law of nature']."[112] Certainly, the kind of extreme nationalism and cultural parochialism espoused by the fascist ideologues of the day must have seemed to Vernadsky to be in strict conflict with his biogeochemical hypothesis that humanity, as a whole, was converging into a geophysical force on a global scale, and, in particular, with his organicist ontology of an intrinsic telos in nature toward such a convergence. For insofar as he considered the activity of organisms—humans included—to be fundamental rather than supplemental to the geospheric conditions of the earth, the very impetus of living matter to harness, store, and release ever-greater amounts of energy, which requires evermore complex and global infrastructures of organization, must therefore be in line with the planet's dynamics of self-organization as a whole. Indeed, it would be artificial—and ultimately self-destructive—on the part of humankind to cease the reproduction of its own conditions for survival. The inevitable winners, in Vernadsky's quasi-Darwinian terminology, are those who understand the creative destructive tendencies of nature and then take it upon themselves to act accordingly. Precisely as the biospheric conditions for life were initially produced through organized patterns of living matter that deepened and strengthened their own conditions in a manner imperceptible to themselves because of nature's unconscious self-organization, so the emergence of the noösphere marks the formation of novel,

112 Ibid., 8.

natural-evolutionary attractors that cannot be broken off by any regressive accidents in human history.

In this sense, with the emergence of the noösphere, the relation of recorded human history to the prehistory of the emergence of *Homo sapiens*, and even more radically, to the deep time of the emergence of the biosphere, becomes evident to humanity for the first time *in* human history. Admittedly, this process of the self-conscious recognition of humankind's production of its own biospheric conditions, and its realization of the necessity of taking upon itself an active role in producing the biosphere for the sake of the maximally vital expression of the planet as a whole, is contingent upon the course of the modern sciences and tied up with technological development. But humankind's production of its own biospheric conditions is by no means an accident or a contingent phenomenon: what was previously an unconscious drive has merely become a self-consciously recognized necessity.[113] Struggles against such natural-evolutionary attractors are thus destined to fail because nature will ensure that even its most radical adversaries will either find their freedom expressed in its productive becoming or else consigned to the dustbin of history. Ideological state apparatuses that either ignore or reject the fundamental unity of humans in nature are, as Vernadsky sees it, either oblivious to, or are outright denying, what the modern sciences have discovered about the workings of the natural world, namely, that all men are unmistakably equal in their ontological subsumption in the earth's self-organization.

§

The inevitability of the transformation of humans along with their transformation of the earth is a theme that looms large in Teilhard de Chardin's work too. If his writings on the natural evolutionary significance of technology can be partly understood as an elaboration upon the implications of Lamarck's — and later

113 Vernadsky, "The Transition from the Biosphere to the Noösphere," 17–19.

Bergson's — proposal of orthogenesis, it should be stressed that they also contain an important paleoanthropological aspect, in that he, drawing upon the work of Le Roy, was interested in the invention and use of technical artifacts and modes of organization primarily as a process of hominization, that is, the transformation of humankind through its use of tools. In 1928, Le Roy had already drawn attention to the hominizing feature that the noösphere, in his view, constituted. Through use, humans develop intimate, habitual, and embodied relations with global environmental networks of machinery and modes of organization, and the noösphere, therefore, does not solely signify humankind's instrumental engineering of its terrestrial lifeworld, but points at least as much to its reliance on novel technological habitats. Put differently, it establishes a dynamic relationship between organism and environment. By extension, Teilhard de Chardin's organicism thus also connects onto a larger, speculative-anthropological project: If technological evolution is bound up with the institutionalization of new modes of existence, then to speculate on where technology might be heading is also to speculate on the future of the human condition. As Teilhard de Chardin argues, the modern technical apparatuses of calculation and communication constitute a spatially and temporally nested "machine which creates, helping to assemble, and to concentrate in the form of an ever more deeply penetrating organism, all the reflective elements upon earth."[114] Even though he did not live long enough to experience our contemporary era of global satellite and telecommunication technology, Teilhard de Chardin already marveled at the way in which media such as radio and television seemed to promise the direct syntonization of brains," and admired the electronic computer's ability to process and communicate information at a pace greater than the speed of human thought, media that, he wrote, "already link us all in a sort of 'etherized' universal consciousness."[115] Impelled

114 Pierre Teilhard de Chardin, *The Future of Man* (New York: HarperCollins, 2008), 162.

115 Ibid.

not only by the analogy but the homology that he saw between zoological and technological evolution, Teilhard de Chardin speculated freely about the implications of an emerging noösphere, as humans themselves were becoming synthesized as novel elements in an artificial platform destined to give rise to an increasingly global awareness, a sublation of self-consciousness that was pulling the within of nature onto a higher order of complexity: "Technology has a role that is biological in the strict sense of the word: it has every right to be included in the scheme of nature. From this point of view, which agrees with that of Bergson, there ceases to be any distinction between the artificial and the natural, between technology and life, since all organisms are the result of invention; if there is any difference, the advantage is on the side of the artificial."[116] In this sense, technology, by its magnitude, complexity, and availability, constitutes environmental devices. It is not only that we as individuals are never separate from the technological milieu within which we dwell, but that our very individuation as such, along with our autonomy, creativity, and desire, are configured by this system.

Although Teilhard de Chardin speculated neither on the genesis nor on the cultural and historical embodiments of self-consciousness, what he did argue was that out of these anonymous beginnings had grown an increasingly refined capacity for comprehending our being on the earth, a historical process during which we have witnessed not only the semantic exchange of different names for this most abstract idea of a self-determining collective, but one that, as the understanding of this collective has morphed in shape, has also seen the negotiation of descriptive and normative content. This is not to deny that there are also new risks involved.[117] Humans leave behind a vast trail of waste and garbage — technical excrements — everywhere they go, and there seems to be no stopping them either. Indeed, humanity's ability to ruin the planet, and to destroy most forms of life — including human life itself — seems to point to the

116 Teilhard de Chardin, *Activation of Energy*, 159.
117 Smil, *The Earth's Biosphere*, 13.

absurdity of conceiving technological evolution as the self-realization of nature under the guise of humankind. It is precisely at this moment, Teilhard de Chardin admits, that relentless acceleration suddenly gives way to hesitation and reflection, and to a sense of disquiet or even terror. Insofar as modernity, because of the emergence of the noösphere, marks a shift in natural evolution, such that we currently face a situation without precedent in the history of the earth, he finds it unsurprising that we suffer from collective disorientation and a fundamental anguish of being. Something terrible is confronting us as we are taken aback by the enormous responsibilities that are being opened up. This is the reason why it is in humankind, Teilhard de Chardin suggests, that we confront, for the first time in natural evolution, the problem of sin. And, for him, it is the unconscious — the abyssal self-organization that precedes spirit's coming to consciousness and recognizing itself in nature — that constitutes its ungrounded foundation. In other words, sin results from our failure to participate in God's creativity and the fulfillment of the universal body that is our terrestrial domain.[118] The task, then, is for humanity to reconcile itself with its assignment, namely, to assume its responsibility as the midwife of the self-directedness of spirit out of nature by taking it upon itself to transform unconscious nature into thinking and foresight.[119]

Describing his experiences as a stretcher-bearer during World War I, in the posthumously published *Writings in Time of War* (1955), Teilhard de Chardin recalled how the infantrymen were drawn into a unity with the sense of belonging to a whole greater than the sum of its parts. At the front, he found that he and his comrades acted with a single soul, and he depicted such moments, whenever one feels the ego dissolved in the immediate contact with others, as mysteries of a profound affinity. Indeed, in many ways, writings on the noösphere during the interwar period reflected as much a deep revulsion against the twentieth century's dialectical turn of Enlightenment into the

118 Teilhard de Chardin, *Vision of the Past*, 74.
119 Teilhard de Chardin, *Phenomenon of Man*, 227–28.

unprecedented destruction of global warfare — countered by a strong faith in human potential and in the progressive nature of science and technology — as it did a strictly scientific concern in the context of the study of the earth. In a final paper before his death, published posthumously in January 1945, Vernadsky wrote with great positivity, even as he beheld the devastation in which at least 20 million Soviets lost their lives:

> The historical process is being radically changed under our very eyes. For the first time in the history of mankind the interests of the masses on the one hand, and the free thought of individuals on the other, determine the course of life of mankind and provide standards for men's ideas of justice. Mankind taken as a whole is becoming a mighty geological force. There arises the problem of the reconstruction of the biosphere in the interests of freely thinking humanity as a single totality. This new state of the biosphere, which we approach without our noticing it, is the noösphere.[120]

Likewise, what Teilhard de Chardin experienced as an infantryman he also sensed as part of a collegiate body of scientists. Modern science, as he recurrently pointed out in his many works, is also a collaborative enterprise. As the sciences seek to systematize their findings, humanity comes to feel itself synthesized into a mutual identity with nature of a higher order. For in the process of coming to know the world, the boundaries of the subject are drawn out and renegotiated. The global effect of humankind's activities on the earth certainly refers to a moment of global crisis, but primarily to the inevitability of a mutation or a new beginning. Contrary to the position of the beautiful soul, bemoaning the current crisis while overlooking how we already are involved in what we deplore,[121] Teilhard de Chardin emphasizes the aspect of self-reflection that he sees present in reason at all times, raising awareness of how we ourselves are deeply

---

120 Vernadsky, "The Biosphere and the Noösphere," 1.
121 Slavoj Žižek, *Living in the End Times* (London: Verso, 2010), 399.

immersed in the current process, but also outlining emerging options to actively contribute to and become part of the inevitable turn. "Man discovers," declares Teilhard, "that he is nothing else than evolution become conscious of itself."[122]

A Will to Power, in the Noösphere Alone

In *Fundamentals of Ecology* (1953), one of the pioneering works for the institutionalization of ecology as a modern scientific discipline, the biologist Eugene Odum, although he enthusiastically celebrated Vernadsky's writings on the biosphere, found its "replacement" with the noösphere to be "dangerous philosophy," since, as he put it, "it is based on the assumption that mankind is now wise enough to understand the results of all his actions."[123] Clarifying the grounds for his reluctance, Odum warned that humankind's power to transform its environment seemed to him to be growing faster than humanity's own understanding of the potential range of its consequences. Technological evolution, he feared, was outrunning scientific explanation, and the noösphere, if anything, should be understood as a symptom of this deeper problem. As technology attains an increasingly global reach in its power to alter nature, the potential outcomes of any single action seem to grow almost exponentially, such that the chance of predicting the actual outcome will either require a revolution in computational capacity or continue shrinking to become almost minuscule. Indeed, Odum worried that although nature is remarkably resilient, the qualitative limits of many ecosystems can now be exceeded by the human action. Later, his younger brother, Howard T. Odum, an eminent systems ecologist in his own right, would go on to caution that the noösphere be understood as a name for the emergence of an unruly global technological system that marked the limits of human ingenuity rather than as a confirmation of our Pro-

122 Teilhard de Chardin, *Phenomenon of Man*, 221.
123 Eugene P. Odum, *Fundamentals of Ecology* (Philadelphia: W. B. Saunders, 1959), 26.

methean ability to foresee the future and plan accordingly: "A noösphere is possible only where and when the power flows of man, or those completely controlled by him, displace those of nature. This kind of dominance over the power of nature is now prevalent in industrialized areas, but these areas survive only because of the purifying stability of the greater areas of the globe not yet so invaded and polluted."[124]

There are obvious similarities between the two Odum brothers in their respective depictions of the noösphere. In both cases, it is an image of technology as by nature inherently ambiguous, and that, in spite of its initial reliance upon human intention, eventually consolidates into a planetary attractor that is at once the result of instrumental manipulation and the advent of a system too encompassing and too complex to remain in human control. Predicated upon the organic insinuation of its self-organizational capacity, this Janus-faced nature of technology, once it starts gaining its own momentum, brings with it a revolution on the level of subjectivity, shattering the shackles of the Cartesian ego. Although technology, in this sense, is not deliberately hostile to human interests, it nevertheless drags humanity along a transformative process of planetary metamorphosis wherein its attractor has already departed from anthropocentric purposiveness and privileges. Such an understanding of technology, as working in accordance with a fundamentally inhuman model of emancipation that subordinates the political economy of human history to the geological economy of deep time, is a central trope in the writings on the noösphere.

For Vernadsky and Teilhard de Chardin, technology cannot be said to be out of control since it constitutes an agency of its own that was never within the strictly instrumental grasp of humankind to begin with. The sense of a loss of control is, in all actuality, rather a sense of a loss of the ego. If, in modernity, technology—including the emergence of modern science, the development of machine tools, and the rise of manufactur-

124 Howard T. Odum, *Environment, Power, and Society* (New York: Wiley, 1970), 244–45.

ing — begins to appear as an instrument for the sake of the conquest of the earth put to the service of humanity, then it is only because its full power is yet to be felt. When he wrote briefly but positively about the "young science of cybernetics,"[125] it is impossible not to think that Teilhard de Chardin saw, already in the 1940s and '50s, the vitalization of technology as a general tendency toward the emergence of increasingly complex systems, irrespective of the material makeup of such systems. Meanwhile, humans find their role in this process to devolve from that of a being who externally controls and steers this dynamic to that of a constructive part in its self-organization. From this perspective, such a vital technological conception appears less as an esoteric critique of instrumental reason on the periphery of the modern imagination than as a serious attempt to make sense of the feedback loops of a technological environment — a cybernetic ecology — too complex to be sufficiently explained within the confines of mechanistic philosophy. Here, the advent of the noösphere marks a threshold in evolution whereby technological systems can no longer be described as mere extensions of biological life since they have become so environmentally distributed, so dispersed across the terrestrial sphere, that organisms themselves are increasingly enrolled as the cogs and gears in their organic machinery of autonomization and reproduction. As modernity comes to its culmination, instrumentalism is destined to give way to a more nuanced understanding of technology as the continuation of terrestrial evolution by other means. In stripping nature of any essential qualities, mechanism reveals a uniformity in nature — immanently connecting the smallest entity to the largest, the oldest to the most recent — and by doing so, ultimately discloses it as a single, evolutionary becoming.

Before the emergence of *Homo sapiens,* natural selection set the course of morphogenesis. After humankind, it is slowly the

---

125 Pierre Teilhard de Chardin, *Man's Place in Nature,* trans. René Hague (New York: Harper & Row, 1966), 110.

power of invention that grasps the evolutionary reins.[126] Importantly, however, even though the word "invention" might seem to suggest that technology plays an instrumental role in evolution, it is by no means that of an instrument operated by a Cartesian subject. From the point of view of Teilhard de Chardin, there is a certain ingenuousness in the Cartesian dualist's confidence that human nature is left unaltered despite the modification by technology of our bodies. The Cartesian imagines that we use technology to modify the material world in our own service, so as to merely "enhance" the mental capacities of the mind, "leaving man free to explore, to create, to think, and to feel."[127] Humankind's essence is then relegated to the spiritual domain, quite distinct from the material aspects that are incorporated into evolution, such that technology is depicted as strictly instrumental, like a designed machine added to a living organism. In this manner, the soul is guaranteed to remain forever human even as the body is cybernetically altered. On the contrary, Teilhard de Chardin stresses continuities on all levels, not only the instrumental parts of ourselves, but our very selves. Similarly, for Vernadsky, life is characterized precisely by its reproduction of itself. All organisms, in his view, seek to act on their environment using whatever devices they have at their disposal. For most organisms, these are natural organs, but for humans it is primarily artificial organs. In this sense, Descartes was not entirely off the mark in stressing the functional similarities between organisms and machines—he just got it backwards. If we, the children of modernity, can be fruitfully described as an unhappy consciousness experiencing ourselves as divided within and against ourselves, it is not because we are the alienated survivors of an Arcadian past, but rather because we are subject to an impending posthumanity.

---

126 Ilia Delio, "Transhumanism or Ultrahumanism? Teilhard de Chardin on Technology, Religion, and Evolution," *Theology & Science* 10, no. 2 (2012): 156.

127 Manfred E. Clynes and Nathan S. Kline, "Cyborgs and Space," in *The Cyborg Handbook*, eds. Chris H. Gray, Heidi J. Figueroa-Sarriera, and Steven Mentor (London: Routledge, 1995), 31.

For as much as the noösphere came to constitute a node in the New Age discourse of the 1960s and '70s, through its portrayal of late modernity in terms of an axial age, it was interestingly retrieved, again, two decades later, as a central idea in the digital cultures of the 1990s, then in an effort to emphasize that the expanding network of technological interconnection enveloping the globe was not something to be questioned but rather embraced as both necessary and inevitable. In fact, in a discourse remarkably similar to the spiritually inclined generation of the late 1960s and early '70s, a slew of futurists argued that the novelty of the digital communication technologies coming to the fore around the turn of the millennium would usher in a cybernetic ecology in which humanity and earth were at last reconciled, and the challenges of the old dualisms of the mechanistic worldview bypassed, as the whole-systems operations of planetary evolution was steering technological development toward unpredictable but ever-more richly complex, diverse, and robust new forms.[128] Although appropriated for what looks at first glance like opposite ends, it is perhaps not so far-fetched, as we have seen, to suggest that the gratification of the desire for the reconciliation of humans and nature in the wake of modernity, both in the case of the New Age movement and in its Promethean counterpart of the 1990s, in fact necessitated a circumvention of the instrumentalism of modern science by means of an exaltation of artifice as poiēsis and an intuitive participation in nature's self-organization. For only in accordance with such an organicist ontology could instrumentalism be set free from its utilitarian constraints of a mere means to become mythologized into an end in itself.

This is to caution, in a characteristically Althusserian fashion, that the noösphere smuggles idealism into materialism by filling the inorganic inwardly with spirit. In a famous 1967 essay responding to the inaugural lecture of Monod, who earlier

128 William H. Bryant, "Whole System, Whole Earth: The Convergence of Technology and Ecology in Twentieth Century American Culture," PhD diss., University of Iowa, 2006, 249–50.

that same year had been appointed professor of molecular biology at the Sorbonne, the French philosopher Louis Althusser criticized precisely his appropriation of Teilhardian concepts, which, Althusser argued, lead Monod's own ontology toward an unacknowledged "mechanistic-spiritualistic" coupling.[129] For by imposing biological mechanisms onto social relations, Althusser pointed out that Monod, "because he does not control the notions he manipulates in the domain of history, because he perceives them to be scientific, whereas they are merely ideological," consequently "perceives only the *intention* of his discourse and not its *objective* effect."[130] The "objective effect" of which Althusser writes is the mystification of social relations through a mythologization that depicts the present mode of organizing production as the outcome of natural-historical necessity. Ascribing a "survival-of-the-fittest" schema onto social relation by means of the natural development of the noösphere, Althusser argues, is to justify class relations, bourgeois politics, and the apparatuses of capital through reification. In other words, it is to give the impression that industrial capitalism has won out over other — all too artificial and thus unsuitable — alternatives because of natural selection and superiority. Even though the ontological flattening of the modern division between nature and spirit portrays itself as a metaphysical corrective to that insufficiently materialist dualism underlying mechanistic philosophy, it ironically remains an idealism in the most fundamental sense of that word, precisely because it inherits the inclusivist imperative that characterized the geosomatism of the ancient Greeks, albeit in an inverted form. Indeed, three years later, in *Chance and Necessity* (1970), Monod would harshly condemn the "intellectual spinelessness" of Teilhard de Chardin's "biological philosophy" in its "systematic truckling" to reestablish "that old animist covenant with nature."[131] But in

129 Louis Althusser, *Philosophy and the Spontaneous Philosophy of the Scientists & Other Essays*, ed. Gregory Elliott, trans. Ben Brewster et al. (London: Verso, 1990), 150.

130 Ibid., 153.

131 Monod, *Chance and Necessity*, 31–32.

his inaugural lecture, the overtones of Teilhard de Chardin's work was palpable, not the least in the way Monod concluded this speech with a grand charge for an "aggressive, in some ways even Nietzschean, ethic," composed of a "will to power, in the noösphere alone."[132] But, as Althusser warned, those who affirm such a will to power are doomed to mix human axiology with independent reality in their circumspect pursuit of absolving humans of culpability for their assertions vis-à-vis objective matters.

132 Jacques Monod, "From Molecular Biology to the Ethics of Knowledge" (trans. Arnold Pomerans), *The Human Context* 1, no. 4 (1969): 336.

4

Mythology in the Space Age

On July 4, 1994, when then Czech president Václav Havel was awarded the Philadelphia Liberty Medal, he delivered an acceptance speech at Independence Hall titled "The Need for Transcendence in the Postmodern World." It was very well received and was subsequently printed in national and international newspapers. In his speech, Havel indicated that there were reasonable grounds for believing that the West was undergoing a transitional period. It seemed to him as if the modern age had exhausted itself. But with modernity crumbling and decaying, something new, still inarticulate, was beginning to take shape out of the rubble. From Havel's point of view, the transition in question was related to the challenge toward — or even, he suggested, "the crisis of" — mechanism as the basis for the modern understanding of nature. Although modernity's complete dependency on the objectification of nature and unconditional faith in the existence of rationally intelligible laws underlay the achievements associated with the modern notion of progress, in effect rendering our physical existence less precarious, at the same time, he argued, it had emptied the world of any spiritual meaning of humanity's own belonging. Today, he declared, experts can explain anything in excruciating detail, yet we understand the structure of our own lives — how it all hangs together — less and less. Modern science, he argued, "produces what amounts to a state of schizophrenia: man as an observer is becoming completely alienated from himself as a being."[1] We live in an age of unprecedented instrumental capability to manipulate nature, yet one without any greater sense of purpose or meaning.

Nevertheless, for Havel, the horizon for this historical break should not be understood as something completely unprecedented. As he pointed out, periods of time when the self-consciousness of entire cultures undergoes paradigmatic shifts have occurred repeatedly throughout history. The same kind of break took place during the Hellenistic period, as the Middle Ages were slowly emerging from the remnants of the classical world, and during the Renaissance, which paved the way for the modern epoch. It is the mixing and blending of cultures that is the distinguishing feature of such a transitional period: the emergence of plurality out of the destruction of old dogmas, and the incorporation and homogenization of previous differences to generate new forms of inclusion and exclusion, integration and fragmentation. This is not to be celebrated or mourned, according to Havel, but merely a fact of historical thought proper. Although it may at first glance appear reasonable to believe that the task of establishing stability can be accomplished through technical means—that is, organizational arrangements, such as political and economic instruments—such efforts are ultimately condemned to failure if they do not grow out of something much more fundamental. The crucial task is, rather, to reconsider the foundation for our being: a reconsideration of our relationship to the earth, from which a shared conception of our generally held values might emerge.[2] If it is to be more than just a slogan, such a foundation must not be taken on faith or left unattended in the hope that it will one day solve itself: that the modern sciences will eventually be able to explain reality in its entirety and that, currently, it is merely a case of prag-

1 Václav Havel, "The Need for Transcendence in the Postmodern World," *The Futurist*, July–August 1995, 47.

2 Ibid., 48.

matically holding out until we get there. Quite the contrary, the foundation for our universality is something that must be continually reinvented, in other words, a task to be carried out by each generation.

Although Havel urged his audience to turn their attention to the crisis of modern science, he held that the potential for the revitalization of meaning and purpose still resided in the sciences, not in a *new* science, but in a science that transcends its own limits so as to understand itself *anew*. The first example discussed by Havel was that of the anthropic principle in cosmology, which he viewed as the return of an antique idea in modern garb: that humankind is more than just an accidental anomaly in an otherwise lifeless and meaningless universe. As a second example of where we find modern science within the purview of myth, Havel mentioned the *Gaia hypothesis,* which depicts the earth as a self-regulating organism greater than the sum of its parts. "According to the Gaia Hypothesis," he maintained, "we are parts of a greater whole. Our destiny is not dependent merely on what we do for ourselves but also on what we do for Gaia as a whole. If we endanger her, she will dispense with us in the interests of a higher value—life itself."[3] Havel was hopeful that these age-old questions, anticipated by cultures and religions around the world, also happened to be the same kinds of questions that arose within scientific inquiry: they relate to the basis of humanity's understanding of itself, of its place in nature, and, ultimately, of nature as such. According to Havel, our hope lies in a revitalized self-understanding, rooted in a renewed understanding of our relationship to the earth, and, by extension, to the cosmos. Somewhat surprisingly for a former political dissident, he even granted that although politicians at international forums may consistently reiterate the universal respect for human rights, such an imperative will remain toothless if it does not derive from a genuinely shared conception of being, and instead that "only someone who submits to the authority of the universal order [...] can genuinely value himself and his

3 Ibid., 49.

neighbors, and thus honor their rights as well."[4] Hence, the Gaia hypothesis seemed to him a promising sign that the most fundamental question of being, far from being relegated to the marginal academic interest of philosophers and theologians, evoked a great deal of controversy and heated debate even within the geoscientific community.

In the third lecture of his 2013 Gifford series, "Facing Gaia: A New Enquiry into Natural Religion," the anthropologist Bruno Latour, when exploring the possibilities for a universal earth-based community in a similar fashion to Havel, again came up against the peculiarities of the Gaia hypothesis. In his lecture, Latour postulated the likelihood that very soon, in the history of science and in the popular imagination, a counter-Galilean paradigm shift will lead us to retell the story of the universe in reverse order: not from the narrow confines of Venice outward to the entirety of the cosmos, as in the case of Galileo, but from the whole of the cosmos inward to the margin of our blue planet.[5] Perhaps, he even suggested, this crucial turning point had already taken place in the fall of 1965, at the Jet Propulsion Lab in Pasadena, California, where James Lovelock, a British atmospheric chemist, inventor, and independent researcher, was drafting a paper together with the philosopher Dian Hitchcock on how to reason on the possibilities for the existence of lifeforms on Mars. In 1965, Lovelock and Hitchcock were both hired by NASA as consultants for its Viking program. This had grown out of the previous and even more ambitious Voyager, which intended to send probes to Mars as precursors to a manned mission scheduled to take place sometime in the 1980s. But as the ambitions of the Apollo Applications Program had to be curbed because of budgetary concerns and a lack of public interest, funding for Voyager was eventually cut in 1968 and the entire mission was canceled three years later. Despite its cancellation, however, the plan behind the Voyager program was carried out

4 Ibid.

5 Bruno Latour, *Facing Gaia: Eight Lectures on the New Climatic Regime*, trans. Catherine Porter (Cambridge: Polity Press, 2017), 75–77.

by Viking in 1975, which by then still had as one of its objectives the task of searching for signatures of microbial life in the soil of Mars. At the time, these experiments worked on the assumption that evidence for life on Mars would be much the same as for life back on earth. One such proposed series of experiments, for instance, involved dispatching what was in effect an automated microbiological laboratory to sample the Martian soil in order to evaluate its suitability for supporting fungi, bacteria, or other microorganisms. The role of Lovelock and Hitchcock in all this was to design an instrument or a method that could be used to detect life on the planet, which was then largely *terra incognita*.[6] In an unexpected turn of events, however, the two of them would soon come to the conclusion that such a device would be completely unnecessary.

Instead, Lovelock and Hitchcock began to argue that the Viking mission would do better to examine the habitability of the Martian atmosphere rather than looking at its soil, theorizing that all life tends to expel waste gases, and, as such, it would be possible to ground an empirical examination of the existence of life on a planet by observing which gases are present. A much more economical solution, in their opinion, was to let earth-based instruments confirm whether the thin atmosphere on Mars was strongly dominated by carbon dioxide, without signs of modification.[7] If that were the case, they argued, the tenuous atmosphere on Mars could be declared chemically dead. As a result, in their search for Martian life, Lovelock and Hitchcock instead found themselves looking back at the conditions for life on earth. Consequently, one is immediately struck by the inverse symmetry between Galileo's and Lovelock's respective gestures of turning earthbound instruments toward the sky, and yet drawing drastically opposed conclusions.[8] In the winter of 1609, when Galileo turned his telescope to the moon, it

6 James E. Lovelock, *Gaia: A New Look at Life on Earth* (Oxford: Oxford University Press, 2000), 1–2.

7 Dian R. Hitchcock et al., "Detecting Planetary Life from Earth," *Science Journal* 3 (1967): 56–67.

8 Latour, *Facing Gaia*, 84.

surprised him how every planet — the earth included — seemed to behave alike. The earth and the moon appeared to Galileo to have the same dignity: both turned around another center. By looking up toward the sky, the indications seemed to him to support the evidence of a fundamental similarity between our own planet and those of other bodies in motion. From there, the world could be vastly expanded. It was from the premise that the cosmos was everywhere the same that the idea of an objective "view from nowhere" could gain some likelihood, insofar as it would allow interchangeable and incorporeal minds to write out the laws of nature by abstracting from any planetary body no other property than the efficient cause of a billiard ball in motion. If the planets and moon were fundamentally determined by the same mechanical laws, then the modern astronomer could abstract from his observations, for when he has seen one object in motion, he has seen them all. As the historian of science Alexandre Koyré put it, Galileo expanded humanity's horizon from the closed world to the infinite universe, as every single "where" was made literally the same as any other — *res extensa* was indeed extensively expanded.[9] However, thanks to the power of the same detached panoptic, Lovelock and Hitchcock, as if by looking down on our terrestrial abode from the heavens, instead saw evidence of a highly peculiar planet. This time, though, the position had been inverted: what they saw was a planet unlike the others.

With their colleagues from Voyager busy devising expensive equipment to be physically transported to and safely landed on the surface of Mars, Lovelock and Hitchcock found themselves in the contrarian position of suggesting that in order to answer the question about life on our red neighbor, the sensible option would be to cancel such plans altogether and instead carry out the rest of the project from where they were, in Pasadena, by directing a considerably less expensive infrared instrument toward Mars to examine whether or not its atmosphere was

9 Alexandre Koyré, *From the Closed World to the Infinite Universe* (Baltimore: Johns Hopkins University Press, 1957).

chemically at the equilibrium state. It would be no less predictable to look at the Martian atmosphere from the earth than to fly all the way there merely to sample the soil, they argued, so the latter option seemed to them but a waste of time and resources — not because they had evidence to suggest otherwise, but since, as they argued, the existing models for life detecting experiments were both vague and unrefined.

Is There Life on Mars?

In the light of his discontent with some of the foundational assumptions of the Viking program's biological experiments, one of the first undertakings carried out by Lovelock, in 1965, was to start over from the beginning by outlining a general physical basis for life-detection and to build a model independent of the physical conditions for life on earth. As Lovelock put it, since one does not ordinarily "look for fish in a desert, nor for cacti on an icecap," it would likewise be a mistake to "look for microorganisms of earth-like habits on Mars."[10] Because of the absence of a formal physical statement to generally describe life, from which a generalizable definition for experimental purposes could be drawn, Lovelock provokingly concluded that "it is not surprising [...] that the proposed experiments in life detection all ask the cautious geocentric question: 'Is there life as we know it?'"[11] The problematic premise of the Viking program, according to Lovelock, had to do with its underestimation of the difficulty in envisaging an alien biochemistry that could still prove useful as a model for experimentation. For that reason alone, he argued, "it would seem pointless and very uneconomic to send a space probe to detect a speculative lifeform."[12] Instead, Lovelock sought inspiration in the physical approach to the recognition of life, which, although no more rigorous than the biochemical,

10 James E. Lovelock, "A Physical Basis for Life Detection Experiments," *Nature* 207, no. 4997 (1965): 569.
11 Ibid., 568.
12 Ibid.

nevertheless promised more in the way of the universality of its application. For him, the universality required for generalization across different habitats was precisely what was missing in the existing scientific literature. Vast amounts of data had been accumulated on every conceivable aspect of living species, but from their outermost to their innermost parts, almost nothing had been said about what life itself was. At best it read like a collection of expert reports, where the chemist would testify to the importance of chemical compositions, and the physicist to the radiation of heat and light. Yet, few seemed willing to propose or even be interested in proposing anything close to a systematic definition. Surprised by the relative lack of a comprehensive definition of life as a physical process, Lovelock nevertheless found inspiration in the works of Erwin Schrödinger, John Desmond Bernal, and Eugene Wigner, who had each made an attempt, and had in the process all arrived at largely the same conclusion: "Life is one member of the class of phenomena which are open or continuous systems able to decrease their internal entropy at the expense of substances or free energy taken in from the environment and subsequently rejected in a degraded form."[13] For the sake of designing life-detection experiments, however, Lovelock knew that this definition was still far too wide-ranging.[14] Initially then, he experienced a great deal of difficulty in convincing his colleagues of the benefits of bringing thermodynamics into the debate. As he recalled, one of the few people who found his suggestion persuasive was Dian Hitchcock.

Even if it was too broad and vague, the inspiration that Lovelock found in thermodynamics would at least point him in the right direction. Together with Hitchcock, whose task it was to compare and assess the logic of the various proposals for extraterrestrial life detection, he would go on to develop one of the most crucial aspects of his reasoning, taking as his starting point the boundary between an organism and its environment. With a

---

13 Bernal, quoted in Lovelock, "A Physical Basis for Life Detection Experiments," 568.

14 Lovelock, "A Physical Basis for Life Detection Experiments," 568.

few exceptions (such as, for instance, William Rubey,[15] an American geologist interested in the origins of the earth's atmosphere and oceans; the physicists Marcel Nicolet and David Bates,[16] who together studied the agronomical influences of molecules, such as methane, water vapor, and ozone, resulting from catalytic reactions; and G.E. Hutchinson,[17] often called "the father of modern ecology" because of his contributions to the study of biogeochemical processes) most geochemists at the time accepted a rigid distinction between the atmosphere and the biosphere, and the regulation of the climatic and chemical configuration of the planet, in turn, was modeled by atmospheric chemists as independent of biotic activity.[18] But thermodynamics, on the contrary, seemed to suggest the importance of the interface, or, system boundary, between the industrial factory of a living system (wherein energy and raw materials are put to use so that entropy is locally reduced) and its surrounding environment, from which the resources flow and where the waste products are later discarded. In this manner, organisms are depicted not only as dependent on certain conditions in their environment for the sake of their subsistence, but also as actively modifying the very same conditions merely through the fact of their presence. Although the actions of individual organisms may be minuscule when observed in isolation, the composite effect of the entire biota may be so great that it leads to changes in the structure of their environment, and to such a degree that an ecosystem, through quantitative changes, may even take on qualitatively different properties and become irreversibly altered as a result.

---

15 William H. Rubey, "Geologic History of Sea Water: An Attempt to State the Problem," *Geological Society of America Bulletin* 62, no. 9 (1951): 1111–48.

16 David R. Bates and Marcel Nicolet, "Atmospheric Hydrogen," *Planetary and Space Science* 13, no. 9 (1965): 905–9.

17 G. Evelyn Hutchinson, "The Biochemistry of the Terrestrial Atmosphere," in *The Earth as a Planet*, ed. Gerard P. Kuiper (Chicago: University of Chicago Press, 1954).

18 Lovelock, *Gaia*, 7, 64–65.

Toward the end of the nineteenth century, the Irish physicist and innovator in fluid mechanics Osborne Reynolds had found that the formation of turbulent eddies and vortices in gases and liquids was contingent upon the rate of flow in relation to local environmental conditions. Following Reynolds, Lovelock hypothesized that there will similarly be an energy gradient between a living system and its environment, and that living systems will propel their environment into disequilibrium. Just as eddies and vortices spontaneously bring themselves into existence once the flow exceeds a critical threshold, living systems require a flux of energy above some minimal value in order to get going—and to keep going. Given the right conditions, then, such systems may begin to organize themselves, and not merely as passive recipients of ecosystem services, but as active producers, playing a crucial role in maintaining the conditions necessary for their own preservation.[19] Only by continually altering the composition of their environment do living systems "stay alive," so to speak. Supposing that extraterrestrial life would also be bound to alter its environment in a similar manner, the internal entropy reduction characteristic of life would spill over onto the Martian environment.[20] "If life is defined as a self-organizing system characterized by an actively sustained low entropy then," Lovelock argued, "viewed from outside each of these boundaries, what lies within is alive in the context of thermodynamics."[21] Put simply, life can be functionally investigated as a dynamic process that is in constant motion, such that its presence is signaled by its effects on the flow of energy needed to sustain it. In the process of staying alive, a self-organizing system produces an outward flux of entropy across its boundary—a bio-signature that can be observed in its environment. If one were to observe such a bio-signature, then one should also come across evidence

19 James E. Lovelock, *The Ages of Gaia: A Biography of Our Living Earth* (New York: W.W. Norton, 1988), 26.

20 James E. Lovelock and C.E. Giffin, "Planetary Atmospheres: Compositional and Other Changes Associated with the Presence of Life," *Advances in the Astronautical Sciences* 25 (1969): 179–93.

21 Lovelock, *Ages of Gaia*, 27.

in the atmosphere that there are lifelike processes on the level of the biosphere without having to rely on an exact or detailed description of what this life is. Instead of searching directly for life—whatever it is—one would do better to look for evidence of its effects on the environment.[22] According to Lovelock and Hitchcock, if there is life on Mars, then the Martian atmosphere is likely to constitute a crucial part of its environment, and its chemical composition should for that reason reflect whether its required energy gradient is present.

Consequently, Lovelock and Hitchcock emphasized that the earth's atmosphere is not an independent and isolated entity that is merely exploited by the biosphere. Rather, the biosphere continuously produces the atmosphere, which, like "a cat's fur, a bird's feather, or the paper of a wasp's nest," amounts to "a biological construction [...] an extension of a living system."[23] Despite the fact that the earth regularly travels around what amounts to an uncontrolled radiant heater—the sun—whose output is erratic to say the least, it is quite significant, as Lovelock and Hitchcock pointed out, that from the origin of life onward, the mean surface temperature of our terrestrial home has varied by at most a few degrees Celsius.[24] Their simple but radical point was that life on earth had not merely adapted to an inert environment, but continuously modified and maintained it in ways that were beneficial for the perpetuation of the biosphere, that is, as a case of self-maintaining feedback. In other words, it might be abstractly conceived of as a cybernetic system.[25] Such a system employs a circular logic, which may appear strange if one has been taught to think in mechanistic terms of linear cause and effect. Rather than the effect of one closed and fixed entity upon another, any insight into the mode of action or performance of a cybernetic system can only be grasped by

22 James E. Lovelock and Dian R. Hitchcock, "Life Detection by Atmospheric Analysis," *Icarus: International Journal of the Solar System* 7, no. 2 (1967): 149–59.

23 Lovelock, *Gaia*, 10.

24 Lovelock and Giffin, "Planetary Atmospheres."

25 Lovelock, *Gaia*, 45.

accounting for its temporal process of operation. For a system to maintain itself, it must continually adapt to its environment, adjusting itself by responding to the feedback of its previous action and so forth. Since its environment is not a passive background but is itself bound up in the ongoing process of operation — both altering and being altered by the system — the organism would not be what it is without the environment, but nor would the environment without the organism. It suggests that the interface between organism and environment is what is crucial for the operational pathways of living systems, in that such a system works to preserve its order, but that it can do so only by an inflow and outflow through its environment.

Although the argument presented by Lovelock and Hitchcock was evidently in conflict with the conventional geochemical understanding of the mid-1960s, after several rejections of their paper they eventually found an open-minded supporter in the American astronomer Carl Sagan, who decided to publish it in *Icarus,* the journal for planetary science he was editing at the time. In the not-so-humble words of Lovelock, this was partly because their proposal to conduct atmospheric analyses proved intimidating to the prevailing scientific consensus, insofar as, considered solely on its merits as a life detection experiment, it was, "if anything, too successful."[26] Here was convincing evidence, they argued, in one comparatively simple test, for a dead planet. For if the earth's atmosphere was so far departed from the state of chemical equilibrium that it could theoretically be observed by an infrared telescope sited as far away as our red neighbor, then why should the reverse not apply to Mars? If the atmospheric disequilibrium on earth advertises the presence of living systems, then its equilibrium on Mars clearly indicates its absence. To their great delight, by using infrared multiplex spectrometry, the wife-and-husband duo of Janine and Pierre Connes would shortly afterward confirm Mars to have an atmosphere dominated by carbon dioxide, which did not significantly

26 Ibid., 7.

depart from the expectations of equilibrium chemistry.[27] As a result, Lovelock and Hitchcock became increasingly convinced that the only feasible explanation for earth's highly improbable atmosphere was that it was being recurrently manipulated at the surface, and that this ongoing process of manipulation was the indication of life they had been searching for all along — general enough to satisfy the criteria of being applicable beyond earth-like constraints, but at the same time specific enough to serve as a working life detection experiment. In striking contrast to that of Mars, they stressed, the earth's atmosphere is in a state of profound chemical disequilibrium, and such a condition can be maintained only by continuous replenishment. At the suggestion of his friend and neighbor, the novelist William Golding, Lovelock named this idea "Gaia," after the goddess of the earth and the primordial mother of all life in Greek mythology.[28]

§

Despite this initial success, however, there remained a long and winding way from the formative years of Lovelock's contribution to extraterrestrial life detection to the point of systematically working out a hypothesis on planetary habitability as an emergent and self-regulatory phenomenon. After having left NASA in 1966, Lovelock continued to work as an independent researcher on the global consequences of air pollution from the combustion of fossil fuels. Committed to furthering his argument that the atmosphere ought to be conceived of "as a component part of the biosphere rather than as a mere environment for life,"[29] it appeared to him that a proper understanding of air pol-

27 Janine Connes and Pierre Connes, "Near-Infrared Planetary Spectra by Fourier Spectroscopy. I. Instruments and Results," *Journal of the Optical Society of America* 56, no. 7 (1966): 896–910.

28 Lynn Margulis, *Symbiotic Planet: A New Look at Evolution* (New York: Basic Books, 1998), 118.

29 James E. Lovelock and Lynn Margulis, "Atmospheric Homeostasis by and for the Biosphere: The Gaia Hypothesis," *Tellus* 26, nos. 1–2 (1974): 2.

lution would require taking into consideration the biospheric feedbacks that might either lessen or intensify the perturbations.

It was at a 1969 conference on the origins of life on earth, in Princeton, New Jersey, that he first presented what would later come to be known as the Gaia hypothesis. Although the reception was a disappointment — his paper attracted at best scant attention — the idea appealed to two of the editors of the conference proceedings, one of whom was the Swedish chemist Lars Gunnar Sillén, an active figure in compiling data on thermodynamic equilibrium, and the other the American evolutionary biologist Lynn Margulis, at the time a junior faculty member at Boston University and one of the foremost proponents for the importance of symbiosis as a driving force in evolution. Lovelock and Margulis met again a year later, in Boston, and began what would turn out to be an intellectually fruitful collaboration. As Margulis recalled, she initially sought out the advice of Lovelock:

> In the early seventies, I was trying to align bacteria by their metabolic pathways. I noticed that all kinds of bacteria produced gases. Oxygen, hydrogen sulfide, carbon dioxide, nitrogen, ammonia — more than thirty different gases are given off by the bacteria whose evolutionary history I was keen to reconstruct. Why did every scientist I asked believe that atmospheric oxygen was a biological product but the other atmospheric gases — nitrogen, methane, sulfur, and so on — were not? "Go talk to Lovelock," at least four different scientists suggested. Lovelock believed that the gases in the atmosphere were biological.[30]

During the late 1960s, as she was studying the structure of cells, Margulis had begun to advance the thesis that organisms primarily evolve through sporadic symbiosis with other organ-

30 Lynn Margulis, "Gaia Is a Tough Bitch," in *The Third Culture: Beyond the Scientific Revolution*, ed. John Brockman (New York: Simon & Schuster, 1995), 139.

isms as opposed to the gradual, Darwinian process of natural selection through the mechanism of individual competition. She took her inspiration from the work of, among others, the Russian botanist Boris Kozo-Polyansky, who in the 1920s had attempted to synthesize the experiments of a number of evolutionary biologists in order to substantiate his theory that "symbiogenesis" — ecological relationships across species or taxa that, over a long period of time, may give rise to new behaviors and structures, including new tissues, organs, species, genera, or even phyla — had constituted the major source of innovation in evolution.[31] Gathering examples of symbiogenesis from previous research on different groups of organisms, Kozo-Polyansky abstracted from these particulars a general principle, from which he speculated that even cells, which back then were considered the elementary units of life, were themselves first and foremost a product of symbiosis and thus more like a cooperative system.

By pursuing this route, Margulis had made herself a fringe figure in the evolutionary biological community. She struggled to even find a single journal agreeing to publish her research. She recalled that her formative paper on the subject, "On the Origins of Mitosing Cells" (1967), was rejected by no fewer than fifteen journals before it was finally published.[32] But Margulis was not deterred. Later, she would credit some of her confidence to an appreciation of the history of biology, which she believed to be crucial for entertaining a greater perspective on matters. Having been encouraged to read widely during her graduate studies, and to not be afraid of diving into the classics in the field, Margulis already knew that, ever since the end of the nineteenth century, biologists had been struck by the similarities between mitochondria and bacteria. Moreover, there were additional parallel examples found in plant cells. Algae, for

31 Lynn Margulis, "Symbiogenesis — A New Principle of Evolution Rediscovery of Boris Mikhaylovich Kozo-Polyansky (1890–1957)," *Paleontological Journal* 44, no. 12 (2010): 1526.

32 Margulis, "Gaia Is a Tough Bitch."

instance, have a second set of bodies—chloroplasts—that are responsible for photosynthesis, which, like mitochondria, bear a remarkable semblance to bacteria in that they have their own DNA, which is separate from the DNA found in the nucleus of the cell. Furthermore, a double membrane surrounds them, which suggests that each was ingested by a primitive host. They also reproduce like bacteria, replicating their own DNA and directing their own division. Her theory was thus that organelles, like the mitochondria, had not been original to the human cell, but had rather come from the outside. It was probably infected, initially, by cellular parasites that fed on the nutrients inside the cell. But eventually, as the amount of oxygen drastically increased in the atmosphere, these parasites became vital for the survival of the cells. Once it became possible to analyze the DNA in the nuclei of human cells, Margulis hypothesized that the cellular DNA would be found to not code for all of the organelles, in effect attesting to their extracellular origin.[33] Less than a decade later, as it became practically feasible to test Margulis's hypothesis, the experiments indicated she had in fact been right. The origin of mitochondria from bacteria, and chloroplasts from cyanobacteria, was confirmed by the biochemists Robert Schwartz and Margeret Dayhoff, and when they published their paper "Origins of Prokaryotes, Eukaryotes, Mitochondria, and Chloroplasts" in 1978, what Margulis had proposed quickly went from being a long-forgotten theory to being generally accepted by the scientific community and finding its way into standard textbook biology.

For the purposes of Margulis's own work then, the autopoietic conception of active self-maintenance implied in the Gaia hypothesis proved particularly suitable for countering the gene-centered, neo-Darwinian account of evolution, which combined Gregor Mendel's genetics with Darwin's account of

---

33 Lynn Margulis, *Origin of Eukaryotic Cells: Evidence and Research Implications for a Theory of the Origin and Evolution of Microbial, Plant, and Animal Cells on the Precambrian Earth* (New Haven: Yale University Press, 1970).

gradual evolution via natural selection. According to one of its most renowned proponents, Richard Dawkins, ecological interactions are subordinated to the gene's desire to be reproduced in the next generation, an urge that triggers the competition for resources.[34] Dawkins proposed the metaphor of "the selfish gene" to express the idea that genes strive for immortality, and that greater units of selection, such as the chromosome, the individual, the family, or the species, are nothing more than vehicles for realizing its goal. The behavior of all living entities is in service of their genes as the primary unit of selection, and it just so happens that the best way for them to survive is in concert with other genes. Natural selection will favor genes that build themselves an organization that is most likely to succeed in safely handing down a large number of replicas to the next generation. Dawkins therefore argued that, metaphorically, genes are selfish. The genes that are passed on are those whose evolutionary consequences serve their own interest in being replicated, and not necessarily those of the organism. Against this, Margulis held that genes cannot be selfish, for the simple reason that they do not operate like an autopoietic system: they do not exhibit the necessary properties, like a membrane, that would allow them to distinguish themselves from their environment.[35] Selfishness requires the maintenance of the functional integrity of a bounded interiority, which can be internally multiple, but is nevertheless operating in distinction to an outside. Meanwhile, Dawkins, even though he opposed the notion that genes are driven by intention, still seemed to imply, in a typically Kantian fashion, that their effects can be described *as if* they were. For Margulis, this was not particular to Dawkins, though, but symptomatic of the whole neo-Darwinian intellectual trajectory of granting primacy to natural selection through competition. Her point is that, in their endeavor to get rid of purpose

34 Richard Dawkins, *The Selfish Gene* (Oxford: Oxford University Press, 1976).

35 Lynn Margulis and Dorion Sagan, *Origins of Sex: Three Billion Years of Genetic Recombination* (New Haven: Yale University Press, 1986), 11.

altogether, the neo-Darwinians threw out the baby with the bathwater: they consequently denied any effort to entertain the question of purpose, which includes the question of the purpose of life.[36] But the concern associated with teleology is not so easily expelled. If Darwin, as Dawkins claims, had done away with the watchmaker of the clockwork that is our universe,[37] this is not the same as to say that he did away with teleology altogether. In fact, teleological notions such as "function" and "design" appear frequently in biology, albeit no longer in the discourse of creationism, but by finding their way into naturalist explanations instead.

Likewise, a fundamental disagreement concerning teleology in nature was at the heart of the reception of Gaia. For a long time, the question of the relationship between self-regulation and sentience—in terms of the capacity for a self-regulating system to act purposefully and with will—would continue to haunt the hypothesis, and like a thorn in the flesh hamper its scientific credibility in the eyes of the wider geoscientific community. In one of the earliest significant publications on the Gaia hypothesis, a 1974 article coauthored by Lovelock and Margulis, they defended "the notion of the biosphere as an active adaptive control system able to maintain the Earth in homeostasis."[38] When the climatologist Stephen Schneider later convened the 1988 American Geophysical Union's Chapman Conference on the topic of the Gaia hypothesis, James Kirchner introduced the idea of a "strong" versus a "weak" Gaia—after which Lovelock was associated for some time with the stronger version, and Margulis with the weak.[39] In her "Gaia Is a Tough Bitch" (1995), Margulis clearly stated where she departed from the idea as it was conceived of by Lovelock, which primarily had to do

36 Margulis and Sagan, *What Is Life?*, 224.

37 Richard Dawkins, *The Blind Watchmaker: Why the Evidence of Evolution Reveals a Universe without Design* (New York: W.W. Norton, 1986).

38 Lovelock and Margulis, "Atmospheric Homeostasis by and for the Biosphere," 3.

39 James W. Kirchner, "The Gaia Hypothesis: Can It Be Tested?," *Review of Geophysics* 27 , no. 2(1989): 227.

with her dislike of the equivalence between earth and organism, since, as she frankly put it, "no organism eats its own waste."[40] According to Margulis, the evolution of multicellular organisms can be generally understood as an expansion of interactive capacities, opening up both the closure of metabolism and the flexibility of the regulatory closure of regeneration—and this removes the crispness of such a demarcation insofar as the export and concomitant dispersion of the organism's manufacture into its environment denotes a loss of internal regulatory organization. In *Acquiring Genomes: A Theory of the Origin of Species* (1992), Margulis and Dorion Sagan argued that "the completely self-contained 'individual' is a myth that needs to be replaced by a more flexible description."[41] Instead, they depicted organisms less as autonomous individuals than as ecosystems, that is, as "communities of bodies exchanging matter, energy and information with each other."[42] In any case, the point is that life does not adapt to a passive physiochemical environment but actively produces and modifies its surroundings. "Gaian regulation, like the physiology of an embryo, is more homeorrhetic than homeostatic," writes Margulis, "in that the internally-organized system regulates around moving, rather than fixed-from-the-outside, setpoints."[43] Although she agreed with the neo-Darwinians that depicting the earth as an organism might be misleading since it cannot be submitted to commonplace evolutionary standards, such as natural selection and random variation, Margulis nevertheless held that, even if Gaia was not to be equated with the living beings studied by the biological sciences, the organicist terminology is nevertheless helpful as long as one radicalizes Kant's original distinction between physics and biology. Having placed organicism into this more general perspective, she thus figured Gaia in terms of the discord-

40 Margulis, "Gaia Is a Tough Bitch," 140.

41 Lynn Margulis and Dorion Sagan, *Acquiring Genomes: A Theory of the Origin of Species* (New York: Basic Books, 2003), 19.

42 Ibid., 23.

43 Lynn Margulis, "Kingdom Animalia: The Zoological Malaise from a Microbial Perspective," *American Zoologist* 30, no. 4 (1990): 866.

ant and far-from-equilibrium dynamics of an open system.[44] In staking out this path, Margulis's writings on Gaia share a certain affinity with the complex systems science that comes out of postwar systems theory, which in many ways also expanded the concept of autopoiesis beyond its modern scientific inception in the domain of biology.

"Autopoiesis," at least as it was coined by the Chilean biologists Humberto Maturana and Francisco Varela, conventionally described the self-organization of cells.[45] For as Kant had already acknowledged in his Third Critique, livings beings, of which cells are arguably the smallest unit, pose a problem to the mechanistic worldview insofar as they continuously select and transform the elements taken from their environmental mediums, and by doing so produce their own continuation and transformation in their production of selective transformation, which is to say that their fundamental processes are recursive and their operations primordially self-referring. In other words, there are features particular to the phenomenon of life whereby it appears as if the design of the whole is the effective cause of the arrangement of its parts, in effect challenging the mechanistic assumption that the whole can be entirely explained by reference to the parts alone. "For in such a product," Kant points out, "nothing is in vain, without an end, or to be ascribed to a blind mechanism of nature."[46] Indeed, the function of organisms presents us with a peculiar circularity, such that we have to assume that the parts are designed or organized according to a certain plan: organs grow and repair themselves, each acting reciprocally as the means and ends of other organs, and moreover, the organism as a whole can organize itself in such a manner that its form endures over time. Certainly, mechanical sys-

44 Margulis, *Symbiotic Planet*, 119.

45 Humberto R. Maturana and Francisco J. Varela, *Autopoiesis and Cognition: The Realization of the Living* (Dordrecht: D. Reidel, 1980). Interestingly, the original 1972 publication, in Spanish, bore the title "De Maquinas y Seres Vivos," which translates into "On Machines and Living Beings."

46 Immanuel Kant, *Critique of Judgement*, ed. Nicholas Walker, trans. James C. Meredith (Oxford: Oxford University Press, 2007), 204.

tems can also be so complicated that we would find a blueprint helpful, but the guiding assumption was that it would at least be possible in principle to understand any such system through reverse engineering. Although this reasoning still seemed to apply to machines during Kant's time, even so, he argued that it did not in fact apply to organisms. Tools require tool-users, but organisms were, on the contrary, seen as systems of organs that appeared to be self-governing and self-steering, that is, organisms behave as if there was mind in nature.

For these very reasons, Kant rejected the idea that the mechanism of physical phenomena alone could account for the function of organisms. But he also denied that we therefore ought to conceive of plants and animals as created by some supernatural rather than natural force. These commitments led him to reintroduce the idea of natural teleology, but with a transcendental ideal spin: if biology is to constitute a part of natural science, then organisms can be viewed neither as divine artifacts produced by some supernatural demiurge nor as mere cogs in the machinery of a clockwork universe. Contrary to the fundamental mechanism of nature, organisms "are the beings that first afford objective reality to the conception of an end, that is, an end of nature and not a practical end. Thus they supply natural science with the basis for a teleology [...] that it would otherwise be absolutely unjustifiable to introduce into that science — seeing that we are unable to perceive a priori the possibility of such a kind of causality."[47] In effect, were it not for reflective judgment and the principle of its functioning — the rational idea of an intrinsic end — the ability to experience beings as alive, and subsequently to study them in the context of the biological sciences, would be impossible. Note the similarity between Hutton's and Kant's reasoning in this circumstance: "All we can say is that if we assume that it is intended that human beings should live on the earth, then at least, those means without which they could not exist as animals, and even, on however low a plane, as rational animals, must also not be absent. But in that case,

47 Ibid.

those natural things that are indispensable for such existence must equally be regarded as natural ends."[48]

Efficient cause alone will be insufficient for explaining certain biological phenomena, and for this reason, teleological judgments still have a role to play in modern science. The organized being is unique in that it works against the mechanistic analogue of the clock, but also against the vitalist analogue of the divine spark: "An organized being is, therefore, not a mere machine. For a machine has solely motive power, whereas an organized being possesses inherent formative power, [...] a self-propagating formative power, which cannot be explained by the capacity of movement alone."[49] The organism is that which is at once means and ends, and it is this—more than any other attribute—that serves as the basis for Kant's distinction between the living and the nonliving. Accordingly, such judgments apply solely to certain beings on the basis of their inner structure as opposed to their existence per se. Nevertheless, as he points out, the idea that nature contains an inherent purposefulness may still be necessary in the regulative sense for making sense of phenomena that, from the perspective of mechanism alone, appear completely baffling.

But to specify what this limit is proved difficult for Kant. Since he asserted that the organism possesses a kind of purposiveness that was not directed from without, the purposiveness of the organism risked becoming identical to the processes of the organism itself, that is, what the organism is comes to equal how it is. Consequently, life becomes ambivalently situated between interiority and exteriority: at once a set of entities "out there" (livings beings) and yet a continuum that connects the "out there" to the "in here," the very principle of life that defines them: "Although the reflective power of judgement, in accordance with its own principle, must assume this purposiveness to be only subjective, that is, relatively to this faculty itself, it still carries with it the concept of a possible objective purposiveness,

48 Ibid., 196.
49 Ibid., 202.

that is, the conformity to law on the part of the things of nature as natural ends."[50]

For once we are equipped with the concept of self-organization, there is suddenly an abundance of entities in the natural world that appear rather lively themselves. Should we not, for instance, think about ant colonies in these terms too, and what about beehives and coral reefs, or even human cities? Are these not self-organizing entities much in the same sense as the organism, and, in that case, what does this mean for the Kantian distinction between life and nonlife? It is in the vein of such a Romantic rejoinder against Kant that the sociologist Niklas Luhmann's social systems theory, one of the most popular extensions of autopoiesis beyond Maturana and Varela's initial application of the term to the biotic domain, could radicalize the concept to denote global systems of technical production and organization too.[51] Similarly, when Margulis and Sagan write that "the biosphere as a whole is autopoietic in the sense that it maintains itself,"[52] such an autopoietic conception of Gaia as a system need not denote a living system per se, but rather a metabiotic system: a self-generating constellation of complex organization that emerges from the interactions of living and nonliving elements, embodying their integrated intermodulations. As an autopoietic system in the metabiotic register, Gaia need not be equated with the organism per se. Rather, Gaia is better understood as participating in the essential quality of organic nature that is the autopoietic form of complex organization, that is, an emergent and recursive form of self-production and self-maintenance within a metabiotic coupling of abiotic and biotic dynamics.[53]

Bruce Clarke has suggested that this divergence in emphasis between Lovelock and Margulis can be illustrated by the devel-

---

50 Ibid., 353.

51 Niklas Luhmann, "Globalization or World Society: How to Conceive of Modern Society?," *International Review of Sociology* 7, no. 1 (1997): 67–79.

52 Margulis and Sagan, *What Is Life?*, 20.

53 Bruce Clarke, *Gaian Systems: Lynn Margulis, Neocybernetics, and the End of the Anthropocene* (Minneapolis: University of Minnesota Press, 2020).

opment of cybernetics. After all, as an engineer and inventor, Lovelock had begun advancing his hypothesis, long before he began collaborating with Margulis, under the sway of first-order cybernetics. Neither did he refer, in his writings on Gaia, to the notion of recursion as popularized by the physicist Heinz von Förster, nor to the concept of autopoiesis as developed by Maturana and Varela. But as Clarke points out, the Gaia hypothesis nevertheless incorporated concepts of second-order cybernetics too:

> Simply put, first-order cybernetics is about control; second-order cybernetics is about autonomy[. …] Unlike a thermostat, Gaia — the biosphere or system of all ecosystems — sets its *own* temperature *by* controlling it[. …] In second-order parlance, Gaia has the operational autonomy of a self-referential system. Second-order cybernetics is aimed in particular, at this characteristic of natural systems where circular recursion *constitutes the system* in the first place[. …] [N]atural systems — both biotic (living) and metabolic (super organic, psychic, or social) — are now described as at once *environmentally* open (in the nonequilibrium-thermodynamic sense) and *operationally* (or organizationally) closed, in that their dynamics are autonomous, that is, self-maintained and self-controlled.[54]

Influenced by Maturana and Varela, Margulis increasingly distanced herself from the metaphor of the thermostat to instead emphasize the autopoietic nature of Gaia. Seen from such an organic perspective, Gaia no longer comes into view as a system of feedback loops that can be mechanically optimized and instrumentally controlled by an all-seeing and disinterested artificer. Rather, it looks more like a body whose operation is

---

54 Bruce Clarke, "Neocybernetics of Gaia: The Emergence of Second-Order Gaia Theory," in *Gaia in Turmoil: Climate Change, Biodepletion, and Earth Ethics in an Age of Crisis*, eds. Eileen Crist and H. Bruce Rinker (Cambridge: MIT Press, 2009), 295–96.

constituted by the coevolutionary interplay between organisms in an environment. This self-organizational notion of Gaia depicts our planet as the evolution of an animated assemblage of parts operating holistically, which means that the components and also the structure of the earth system are conditional upon natural historical trajectories of creation and destruction over time, as opposed to some programmed set of criteria that could be anticipated ahead of its formation. In fact, even Lovelock himself emphasized this point in his reasoning about the influence of cyanobacteria on the early composition of the atmosphere and the adaptation of other organisms to their alteration of the terrestrial environment. The fact that many organisms today require an abundantly oxygenated atmosphere to survive is not the result of a preordained feedback loop inserted into the earth-machine by a benevolent Creator, but rather because life, during the long haul of deep time, eventually found a way to turn this change in the environment into an evolutionary advantage—only those organisms that managed to turn what was initially a deadly poison to most lifeforms into an accelerator of their metabolism could flourish. Oxygen, then, is not simply bestowed upon the earth as a given part of its terrestrial environment, but is continually produced and maintained, to this day, by the proliferation of organisms and their activities.[55]

In order to attend to the nuanced difference in emphasis present in Lovelock's and Margulis's respective influences upon the formation of a Gaian view of the earth, we will therefore do well to follow Clarke's advice and situate them in the broader intellectual historical context of postwar cybernetics and the accompanying space race between the two superpowers.

Gaia, the Goddess of Cyborgs

Rapid developments in rocketry, computation, and materials science connected to the postwar race to space, along with the Cold War threat of nuclear annihilation, mark a crucial step in

55 Lovelock, *The Ages of Gaia*, 71–73, 114–15.

the historical emergence of a global earth system perspective. Arguably, the space age was inaugurated with the launch of the Sputnik 1 satellite in 1957, and by the time the first images of our blue marble arrived from outer space, in 1965, humanity had already amassed enough weapons of mass destruction to lay waste to the entire biosphere. In its wake, the prominence of existential concern behind theories of military and civilian risk assessment, such as the idea of a "nuclear winter,"[56] would constitute precursors to global environmental change as a scientific object of investigation, but these early predictions entailed the threat of a new ice age rather than that of global warming. In 1974, media theorist Marshall McLuhan thus linked the intellectual roots of the twentieth-century paradigm of ecology to the historical circumstance of spaceflight capability:

> Perhaps the largest conceivable revolution in information occurred on October 17, 1957, when Sputnik created a new environment for the planet. For the first time the natural world was completely enclosed in a man-made container. At the moment that the earth went inside this new artifact, Nature ended and Ecology was born. "Ecological" thinking became inevitable as soon as the planet moved up into the *status of a work of art.*[57]

In this passage, McLuhan draws out a topological inversion that anticipates the thesis of Jacques Ellul's *The Technological System* (1977): humans can no longer dwell innocently within the padded walls of their biospheric cornucopia, for the complete encircling of the earth by human artifice marks the death knell of the idea of a pristine nature unspoiled by anthropogenic interference. Hence, the exclusion of nature from the artificial

---

56 Paul R. Ehrlich et al., "Long-Term Biological Consequences of Nuclear War," *Science* 222, no. 1 (1983): 1293–300.

57 Marshall McLuhan, "At the Moment of Sputnik the Planet Became a Global Theater in Which There Are No Spectators but Only Actors," *Journal of Communication* 24, no. 1 (1974): 49 (my emphasis). Sputnik was launched on October 4, 1957.

dimension of human praxis — such as representations, history, technology, and art — suddenly turns into its complete technical inclusion (or "enclosure," as it were) within an artificial container, one that, in McLuhan's play with metaphor, is represented both by Sputnik's shape and by its orbit. In terms of this all-encompassing world picture, the era of ecosystems science marks the end of nature, and ecology, if we are to believe McLuhan, ought rather to be understood as a discipline that attends to artificial concerns of active intervention into and participation in the production of nature, as opposed to some immaculate domain ontologically distinct from humankind's activities. Alluding both to Shakespeare and military terminology, the globe, in McLuhan's own words, becomes a theater. That there are — as the title of his article suggested — no spectators but only actors, again points to the centrality of immanence: such a technological enframing allows for no outside, nothing that has not already been brought into the workings of the self-organizing earth system, but also, to the aforementioned enactivism that underpins the epistemological assumptions of ecology as a branch of the modern sciences. Not by coincidence, Sputnik itself is also enrolled by McLuhan as a symbol of the Cold War. The notorious shock that accompanied its appearance for the Western world in the context of the arms race is associated with a historical situation characterized by the external boundaries and internal paranoia of political blocs — an intellectual climate that was foundational for cybernetic systems thinking in its fascination with operational closure for the sake of complete control, such that every sphere of life may become a planned, calculated, and organized routine to be executed with maximum efficiency and minimal risk to a military's resources.[58]

In *The Technological System*, Ellul proposed that the increasing systematization of the natural world, during the latter half of the twentieth century, into an organic whole, was to a large

58 Christoph Neubert and Serjoscha Wiemer, "Rewriting the Matrix of Life: Biomedia between Ecological Crisis and Playful Actions," *communications +1* 3, no. 1 (2014): 3–4.

degree a product of the computer's ability to process data.[59] "Technology has the apparatus for allowing the flexibility of the whole," he argued, "namely, the computer. The computer permits the shift from a formal and institutional organization to a relationship by means of information and the dynamic structure according to flows of information."[60] The computer holds the system together, wedding different technologies—technologies that, when operating in concert, provide the means necessary for producing such a totalizing worldview. An important context for the proliferation of the idea of an earth system is thus the association of systems science, as a transdisciplinary approach, with the military-industrial complex. Scholarship on the early history of global environmental change places it firmly within the narrative of Cold War geopolitics, highlighting how the centrality of objectives such as "surveillance" and "control" still echoes the militarized discourse of its intellectual historical roots.[61] The science studies scholar Paul Edwards has compared the notion of a "closed world" underpinning the belief in the planet as something fully manageable with that of a "closed system" in early cybernetics.[62] During the Cold War, there was a growing demand for such a systematic approach to understand the function of the global environment in order to predict and evaluate risks, such as the possible implications of nuclear war or the military detection of atomic test sites by investigating the circulation of isotopes.[63] Similarly, the historian Ronald Doel has traced present-day modes of investigation in global envi-

---

59 Jacques Ellul, *The Technological System*, trans. Joachim Neugroschel (New York: Continuum, 1980), 92.

60 Ibid., 111.

61 Chunglin Kwa, "Modelling Technologies of Control," *Science as Culture* 4, no. 3 (1994): 363–91, and James R. Fleming, *Fixing the Sky: The Checkered History of Weather and Climate Control* (New York: Columbia University Press, 2010).

62 Paul N. Edwards, *The Closed World Computers and the Politics of Discourse in Cold War America* (Cambridge: MIT Press, 1997).

63 Ronald E. Doel, "Constituting the Postwar Earth Sciences: The Military's Influence on the Environmental Sciences in the USA after 1945," *Social Studies of Science* 33, no. 5 (2003): 635–66, and Michael A. Dennis, "Earthly

ronmental change research to the methodological orientations developed in military institutions, such as the geophysics programs at the Woods Hole Oceanographic Institution, Lamont Geological Observatory, and Scripps Institution of Oceanography, during the 1950s, which at the time were all considered novel approaches to the study of the earth.[64] Studies of the arctic atmosphere for the purpose of using and defending against intercontinental ballistic missiles, and, among other things, the involvement of various branches of military intelligence in mapping ocean seafloors and global water circulation for the sake of maritime warfare, have also been raised as examples of the intersection between military strategy and systems science.[65] In this global struggle, the idealization of a completely computerized and predictable nature was advanced as a means to manage the earth as a battlefield, controlled with the help of state-of-the-art technology for surveillance and computation.

The idea of a global observation system has similarly been interpreted in the context of the Cold War tensions, to which environmental concerns were later applied as a means of continuing to promote investment in expensive and large-scale infrastructures even after the fall of the Iron Curtain.[66] Given the state of earth science in the late 1950s, a number of instruments for computing, sensing, and measuring the global environment had been made available from the adoption of military technol-

---

Matters: On the Cold War and the Earth Sciences," *Social Studies of Science* 33, no. 5 (2003): 809–19.

64 Doel, "Constituting the Postwar Earth Sciences."

65 Matthias Heymann et al., "Exploring Greenland: Science and Technology in Cold War Settings," *Scientia Canadensis* 33, no. 2 (2010): 11–42. See also Ola Uhrqvist, "Seeing and Knowing the Earth as a System: An Effective History of Global Environmental Change Research as Scientific and Political Practice," PhD diss., Linköping University, 2014.

66 W. Henry Lambright, "The Political Construction of Space Satellite Technology," *Science, Technology & Human Values* 19, no. 1 (1994): 47–69; Karen T. Litfin, *Ozone Discourses: Science and Politics in Global Environmental Cooperation* (New York: Columbia University Press, 1994); and Kristine C. Harper, *Weather by the Numbers: The Genesis of Modern Meteorology* (Cambridge: MIT Press, 2008).

ogy, which allowed for a scope of investigation entirely without historical precedent—spectroscopes, cosmic ray recorders, and radiosonde balloons made the upper atmosphere available for detailed exploration, while the electric controllable computer facilitated the analysis of vast amounts of information, which had previously been unfeasible to compute and had thus set a limit on the size of datasets.[67] Perhaps the most decisive technology, though, was the rocket. It was only with the post–World War II advances in rocketry that the exploration of space became a real possibility, and neither the Soviets nor the Americans delayed in sending satellites into the earth's orbit. The period during the late 1960s and early 1970s saw further technological developments in space exploration, new earth observation technologies, and the development of computers capable of handling and storing larger sets of data and running calculations at unprecedented rates—two-dimensional maps could now be complemented by computer simulations. Developments pertaining to the computer fundamentally shifted epistemic practices, such as representation, visualization, communication, and simulation, and in effect reshaped the production of knowledge. Computational methods allowed earth scientists to redefine their research problems in line with an entirely new mode of experience: the analysis of complex systems came to involve partial differential equations that could not have been calculated otherwise, or that would have taken unsustainable efforts to pursue without the computational technology of the postwar era. Further expanding the domain accessible to quantification, these technologies opened a window of opportunity for the scientific investigation of the earth. Yet, they also promoted a particular methodology suited to a systems-based ontology.[68]

67 In 1922, the British mathematician and physicist Lewis Fry Richardson famously estimated that effective numerical weather prediction would take no fewer than 64,000 "human computers"; see Richardson, *Weather Prediction by Numerical Process* (Cambridge: Cambridge University Press, 1922), 219.

68 William J. Kaufmann and Larry L. Smarr, *Supercomputing and the Transformation of Science* (New York: Scientific American Library, 1993). See

Alongside access to new global observational instruments, the acceleration in computational power thus instigated a discursive imperative for earth scientists to consider the holistic interaction between geospheres, which, in turn, encouraged the development of models to describe and predict their dynamical behavior. An early example is the American computer engineer Jay Wright Forrester's pioneering work on "world dynamics" in his World2 model,[69] which laid the groundwork for the simulations in the Club of Rome's report *The Limits to Growth* (1972), and which would later provide the impetus for Lovelock's own attempt to formalize the Gaia hypothesis in what came to be called the *Daisyworld model.*

However, Daisyworld was designed with a distinctly different objective in mind, namely, with the intention of refuting the claim that there was some religious or mystical aspect to Lovelock's postulation that the entire earth exhibits homeostatic properties equivalent to those of a living organism, a concern that had already been raised by Kant in his effort to distance modern science from natural teleology. For the Gaia hypothesis had in fact attracted a substantial amount of criticism from a number of evolutionary biologists, among them Dawkins, who held that planetary-scale thermoregulation would be strictly impossible without also extending the mechanism of natural selection to the cosmic scale, and moreover that organisms acting in concert would imply foresight and planning, which runs contrary to the scientific consensus. Ford Doolittle, who rejected the idea of planetary self-regulation on similar grounds

---

also Sarah E. Cornell et al., "Earth System Science and Society: A Focus on the Anthroposphere," in *Understanding the Earth System: Global Change Science for Application*, eds. Sarah E. Cornell et al. (Cambridge: Cambridge University Press, 2012), 1–38.

69 Even Forrester himself started his career as an electrical engineer at MIT during World War II, where he worked to develop servomechanisms for radar antennas and gun mounts. In the 1950s, as he turned from military environments to economic management, Forrester's work nevertheless retained its specific methodological approach to modeling — the rationale of technical and social engineering — and continued to draw upon a very specific technical medium: the computer.

to Dawkins, said that the Gaia hypothesis appeared to entail a "secret consensus" among the biota, and thus some sort of global-scale intention or planetary-wide consciousness.[70] But Lovelock disagreed wholeheartedly. In the preface to the first edition of *Gaia: A New Look at Life on Earth* (1979), he had already warned the reader that

> occasionally it is difficult, without excessive circumlocution, to avoid talking of Gaia as if she were known to be sentient. This is meant no more seriously than is the appellation "she" when given to a ship by those who sail in her, as a recognition that even pieces of wood and metal when specifically designed and assembled may achieve a composite identity with its own characteristic signature, as distinct from being the mere sum of its parts.[71]

In accordance with Lovelock's view, Gaia was to be understood as a cybernetic system with the capacity for self-regulation, and therefore no consciousness or overriding intention was needed. It rather displayed characteristics of such biological systems as beehives, whereby the parts of the system themselves secure the requirements for their own endurance in their ongoing interaction instead of according to some prefixed blueprint. "Gaia is best thought of as a superorganism," states Lovelock. "These are bounded systems made up partly from living organisms and partly from nonliving structural material. A bee's nest is a superorganism, and like the superorganism, Gaia, it has the capacity to regulate its temperature."[72] Quite different, then, from Doolittle's strawman.

Describing the earth as an adaptive control system with feedback loops that maintain homeostasis, the cybernetic jargon

70 Richard Dawkins, *The Extended Phenotype: The Gene as the Unit of Selection* (Oxford: Oxford University Press, 1982), 234–37, and W. Ford Doolittle, "Is Nature Really Motherly?," *Co-Evolution Quarterly* (Spring 1981): 58–63.

71 Lovelock, *Gaia*, ix–x.

72 Lovelock, *Ages of Gaia*, 15.

of Lovelock's writing is unmistakable. In order to avoid confusion, it is important to note that Lovelock defined the earth as a superorganism only because biological organisms themselves had already been cybernetically described as any other system — whether natural or artificial — that operates through the assemblage of interactive parts. "There is little doubt that living things are elaborate contrivances," Lovelock and Margulis wrote in one of their first coauthored papers, adding that "life as a phenomenon might therefore be considered in the context of those applied physical sciences that grew up to explain inventions and contrivances, namely thermodynamics, cybernetics, and information theory."[73] The organic, for Lovelock, was not something peculiar to certain biological beings, but rather applied to all entities in the world that could be said to be in some sense "adaptive." In chapter 4 of *Gaia: A New Look at Life on Earth,* titled simply "Cybernetics," Lovelock extended the meaning of the word "organism" beyond the domain of biology by making use of the ontologically flat concept of system, leading him to analogize the complexity of Gaia to the logistical operation of a business: "Whether we are considering a simple electric oven, a chain of retail shops monitored by a computer, a sleeping cat, an ecosystem, or Gaia herself, so long as we are considering something which is adaptive, capable of harvesting information and of storing experience and knowledge, then its study is a matter of cybernetics and what is studied can be called a 'system.'"[74]

It was for this particular reason that he, just like Hutton, persistently switched back and forth between body and machine metaphors. What Lovelock was after was a transdisciplinary approach that could erase the border between subjects and objects that Dawkins and Doolittle were so keen to patrol. Although he called Gaia a superorganism, Lovelock simultaneously described its function in terms of an immense computer with exceptionally long processing cycles, depicting the

73 Lovelock and Margulis, "Atmospheric Homeostasis by and for the Biosphere," 3.

74 Lovelock, *Gaia,* 57.

genomes of its lifeforms as its most powerful memory bank: "By transmitting coded messages in the genetic material of living cells, life acts as a repeater, with each generation restoring and renewing the message of the specifications of the chemistry of early Earth."[75] But since it does not possess a central processor, Gaia clearly does not adhere to the understanding of the stored-program computer as laid out by the Von Neumann architecture in the 1940s. On the contrary, its information processes were conceptualized to emerge from an interplay of connected yet not fully integrated components, that is, without anything like a kybernētēs — a "captain" or "helmsman" — to steer and guide the process according to a predetermined intention. It is this peculiar characteristic that makes something organic, in Lovelock's view, and not the question of whether it is essentially biological as opposed to technological, or natural as opposed to artificial. In the sense that it also consisted of coupled differential equations that sought to represent complex feedback loops, Daisyworld was thus strikingly similar in structure to Forrester's models of the dynamics of urban planning and corporate management — unsurprisingly so, since both in turn drew upon a methodology that had emerged directly out of the development of cybernetics in the late 1940s and early 1950s.

§

In the second half of the twentieth century, self-organization returned to the forefront of a plethora of disciplines through a radicalized version of Norbert Wiener's cybernetics. In the work of Förster, Maturana and Varela, and the anthropologists Gregory Bateson and Margaret Mead, and also in that of Luhmann, among others, the idea of self-organization was rejuvenated through careful analyses of self-referential loops. Ever since its inception in the late 1940s, the aim of cybernetics had been to provide a general principle of communication. In fact, Wiener himself defined cybernetics as precisely "the science of

75 Lovelock, *Ages of Gaia*, 164.

communication and control," and to underscore the irrelevance of the nature of the communicators involved, he added wording to the subtitle to his 1948 magnum opus: "in the animal and machine."[76] In other words, one of its basic features was that it conceived of the relationship between animals and machines as not merely analogous but homologous. In *Cybernetics* (1948), Wiener remarked that "our inner economy must contain an assembly of thermostats, automatic hydrogen-ion-concentration controls, governors, and the like, which would be adequate for a great chemical plant. These are what we know collectively as our homeostatic mechanism."[77] Accordingly, he thought that, with a sufficiently detailed description of the function of self-organizing systems, it would be possible to construct machines with the same behavior as animals. Contrary to Lovelock's strategy, the architects of cybernetics sought to bypass the confusions and disagreements about the relationship between life and the second law of thermodynamics by turning directly to engineering, or, more specifically, to the design and construction of machines with those very properties that had previously been thought to be exclusive to animals.[78]

In fact, producing not only new concepts, but first and foremost new machines, cybernetics was not simply a theoretical movement. Despite its terminological abstraction, which partly explains the spread of its popularity across disciplinary boundaries, the origins of cybernetics, and its early successes, were clearly technical — it was largely a language that developed out of the design of automatic controllers, telecommunications systems, computers, and other informational networks. Many of those who would later form the movement's core group of intellectuals had been tasked during World War II with working on new weapons systems, radar, and the kind of computational

76 Norbert Wiener, *Cybernetics: Or Control and Communication in the Animal and the Machine* (Cambridge: MIT Press, 1948).

77 Ibid., 111.

78 Evelyn F. Keller, "Organisms, Machines, and Thunderstorms: A History of Self-Organization, Part One," *Historical Studies in the Natural Sciences* 38, no. 1 (2008): 47.

machines that would eventually lead to the electronic stored-program computer. But just as importantly, others were trained in neurophysiology and evolutionary biology, which gave the movement a double perspective — from the technological and the biological domain — that they then sought to connect somewhere in the middle. Since the community of mechanical engineers were largely interested in cybernetics as a novel approach to thinking about machines — as self-regulating systems maintaining their stability through feedback loops — and since the community of cognitive scientists and biologists were concerned with understanding the mechanisms of organisms exhibiting such capacities, it was thought that a conjoint, transdisciplinary study of the function of organized systems was both necessary and inevitable.

Its theoretical development, then, went far beyond the initial technicalities of engineering to cross-pollinate with studies of informational flows, complex systems, and negative and positive feedback, adding the notions of equilibrium and homeostasis, in effect formalizing quite abstract and general ideas. Bateson explained: "The ideas [of cybernetics] were generated in many places: in Vienna by Bertalanffy, in Harvard by Wiener, in Princeton by von Neumann, in Bell Telephone labs by Shannon, in Cambridge by Craik, and so on. All these separate developments dealt with communicational problems, especially with the problem of what sort of thing is an organized system."[79] But whereas early cybernetics was characterized by an interest in the similarities between autonomous living systems and machines, founded on a fascination with new control and computer technologies for the possibilities of a system designer to engineer and steer the function of systems, the epistemological assumptions underlying its mechanistic models were eventually called into question and attention switched to the systemic nature of the designer itself. Importantly, even for Wiener, cybernet-

79 Gregory Bateson, *Steps to an Ecology of Mind: Collected Essays in Anthropology, Psychiatry, Evolution, and Epistemology* (London: Northvale, 1987), 480.

ics had given rise to new questions about human nature. For if human behavior can be duplicated by machines, then how is one to distinguish humans from other self-organizing systems? Historian of science Peter Galison argued:

> On the mechanized battlefield, the enemy was [...] so merged with machinery that (his) human-nonhuman status was blurred. In fighting this cybernetic enemy, Wiener and his team began to conceive of the Allied antiaircraft operators as resembling the foe, and it was a short step from this elision of the human and the nonhuman in the ally to a blurring of the human-machine boundary in general. The servomechanical enemy became, in the cybernetic vision of the 1940s, the prototype for human physiology and, ultimately, for all of human nature.[80]

Admittedly, retaining his emphasis on information and communication, Wiener argued that, in comparison to other beings, only humans are obsessive in their determination to communicate. This was certainly not a satisfactory criterion for the distinction between human and animal, or human and machine, but it did demonstrate that, right from the outset, and in spite of Wiener's own efforts to the contrary, cybernetics played a central role in the erasure of the modern conception of the human subject.

Over time, the cyberneticists thus became increasingly preoccupied with problems involved in conceiving system interactions as a linear, one-to-one proportionality between input and output. Less than a decade after Wiener's publication of the foundational *Cybernetics,* William Ross Ashby challenged the primacy of human communication by connecting it to the flat ontology of Claude Shannon's information theory. And Bateson, in a similar vein, sought to demonstrate how the subject matter of cybernetics extended across natural and social scientific reg-

80 Peter Galison, "The Ontology of the Enemy: Norbert Wiener and the Cybernetic Vision," *Critical Inquiry* 21, no. 1 (1994): 233.

isters by focusing on the purely propositional aspects of communication, thereby further undermining human exceptionalism. As a result of the rapid expansion of cybernetics during the postwar era, Martin Heidegger argued that cybernetics had become a new "fundamental science," typified by the kind of ontological leveling that, in its effort to subdue the entirety of being to the standard of scientific objectivity, ultimately reduces the human to nothing more than an intricate contraption, or to merely a gear in the great machine of the planetary heat engine.[81] Insofar as it seeks to reduce nature to logical or mathematical models — and the question of being to nothing but a uniform mechanism of patterns of communication — in an effort to govern and steer, cybernetics, from Heidegger's point of view, figured as a manifestation of the modern instrumentalist desire for full control.

But although the aim of the cognitivists of the cybernetic movement was, as feared by Heidegger, the formalization of thinking into that of a computational process, it does not follow that it would therefore be contradictory to argue that it also led to the vitalization of the machine. On the one hand, Wiener's science of communication and control represented for Heidegger the culmination of the instrumental reduction of nature into that of a slave to the purposes of humans, whereby every being is disclosed as nothing but an object of their will, fashioned as a function of their needs and desires. Yet, precisely because of the functionalist approach of early cybernetics, which studied systems as passive and external processes that could be freely observed and manipulated, the discipline slowly began to undermine itself from within as its practitioners came to realize that such an understanding suffered from a blind spot, which meant that it remained oblivious to one of the most foundational questions that a study of cybernetic systems must examine, namely, the system interactions between the observer and

81 Martin Heidegger, "The End of Philosophy and the Task of Thinking," in *Basic Writings*, ed. David F. Krell, trans. Joan Stambaugh (New York: HarperCollins, 1978), 434.

the observed implied within the very act of observation. It was argued that cybernetics, if it is to be genuinely systematic, cannot take the observer as an exception to its conceptual framework, but ought to incorporate it into its study as yet another system to be observed. From a Heideggerian point of view, it should thus come as no surprise that, in its utter determination to render the entire earth "for use," cybernetics eventually put the autonomous, Cartesian subject to death, which explains why many of the deconstructionists of the concept of the human have since celebrated and borrowed from cybernetics the ammunition for their own assault. And there were certainly some good reasons why the various projects involved in a deconstruction of metaphysical humanism found in cybernetics a noteworthy ally.[82] To the question of whether cybernetics was the height of metaphysical humanism or if it rather represented the withering away of the Cartesian cogito altogether, the answer, as suggested by the philosopher Jean-Pierre Dupuy, may in fact be "both at once." Because in order for the cybernetic regulation of life to become possible, it was first necessary for it to be reduced to the status of an object: to order, secure, develop, and optimize life is to render it present and available to the disciplinary technologies that are to govern its function. But at the same time, no "lowering down" can occur without a simultaneous "raising up," and vice versa. Consequently, cognition without a subject—indeed,

82 Several scholars have remarked upon the affinities of the ontological trajectories from first- to second-order cybernetics and from French structuralism to poststructuralism. See Niklas Luhmann, "Deconstruction as Second-Order Observing," *New Literary History* 24, no. 4 (1993): 763–82; Cary Wolfe, "In Search of Posthumanist Theory: The Second-Order Cybernetics of Maturana and Varela," in *Observing Complexity: Systems Theory and Postmodernity*, eds. Cary Wolfe and William Rasch (Minneapolis: University of Minnesota Press, 2000): 163–96; Céline Lafontaine, "The Cybernetic Matrix of 'French Theory,'" *Theory, Culture & Society* 24, no. 5 (2007): 27–46; Cary Wolfe, *What Is Posthumanism?* (Minneapolis: University of Minnesota Press, 2009); and Jean-Pierre Dupuy, "Cybernetics Is an Antihumanism: Technoscience and the Rebellion against the Human Condition," in *French Philosophy of Technology: Classical Readings and Contemporary Approaches*, eds. Sascha Loeve, Xavier Guchet, and Bernadette Bensaude-Vincent (Berlin: Springer, 2018), 139–56.

cognition without mental content—turned out to be just the improbable arrangement that cybernetics appeared to demand in order to remain consistent. Hence its efforts to render symbolic thought a product of unconscious structures of physical systems rather than peculiar to individual brains, as if operating in the background, or in the environment of human subjects, who are then rendered no more than a kind of afterthought. For the cyberneticists, "it thinks" was destined to take the place of the Cartesian "I think."[83]

In this manner, cybernetics self-reflexively turned to the role of the observer in the process of observation, and in an effort to include itself among and within the systems observed, it abandoned the modern dichotomy between subject and object. In other words, it got caught up in the circular effort of describing the describer, observing the observer, and pursuing a cybernetics of cybernetics—simply put, cybernetics became "second-order cybernetics." Although the latter marked a shift away from the early cybernetics of Wiener in the sense that it switched attention from the communication between coupled systems to the recursive complexities of communication itself, it nevertheless retained, and importantly so, the functionalist focus on the form of behavior rather than the building blocks of phenomena, that is, on what systems do and how they are observed. Whether natural or artificial, beings were conceptualized as semiautonomous systems coupled with their environment and to other systems. Whatever system is observed, then, will to some degree always owe its existence to other systems within its environment. From such a perspective, autonomy can no longer be solitary, but is instead rethought in terms of operational self-reference: it alters the emphasis of observation and description from that of an already demarcated agent to a process of systems interactions within which agency can emerge, or from the action of an autonomous self to a subjectivism without a clearly demarcated selfhood. Along with this move, the self-

83 Dupuy, "Cybernetics Is an Antihumanism," 144.

present unity of the Cartesian ego in effect gives way to the continual reproduction of certain constellations of beings.

§

John Locke, William Irwin Thompson notes, is often said to have articulated the modern foundation for the enframing of nature into a resource pool of raw materials by defining "property" as that which the labor of humans has transformed from out of the inert and untouched state that nature leaves its products in, such that the natural world passively exists over yonder, merely awaiting a subject to impose its will upon it. If this is the case, then, as Thompson argues, the work of Bateson offers an inversion of the Lockean understanding of nature, so as to view it not as an aprioristic blank slate or background condition, but on the contrary as "unconscious Mind, or Gaia,"[84] which gradually comes into self-consciousness by way of its own laboring upon itself. In Thompson's Gaian enrollment of Bateson's corpus, mind does not reside in the external realm of the *res cogitans* but is rather understood to be one with the living earth, such that its diversity of beings is not in the last instance determined by a difference in quality but by the degree of complexity associated with them — humans being part of this diversity of beings. One of the most central and recurring themes in the otherwise wide-ranging work of Bateson is precisely the dispersal of the self itself into its environment. Although many of the early cyberneticists were in fact cognitive scientists of one sort or another, Bateson was arguably among the first of them to acknowledge that it is the patterns of organization in systems, and their relational symmetry, that is indicative of mind generally defined — again, regardless of whether these systems are natural or artificial.

Pointing to the work of its formative figures, such Wiener and Warren McCulloch, Bateson made it his mission to maintain that, to further the cybernetic agenda, it was precisely mind

84 William I. Thompson, *Pacific Shift* (San Francisco: Sierra Club, 1985), 177.

that had to be examined in more detail, which eventually led him to begin formulating a "cybernetic epistemology" to cover the criteria that were to become crucial to his later ecological philosophy. As a matter of fact, he even considered cybernetics itself to be epistemology, but with the proviso that "epistemology" does not merely denote the Kantian problem of knowing the outside world, but also knowing the self as part of the said world. Proceeding from the work of McCulloch in particular,[85] Bateson was led to the conclusion that epistemology, rather than the question of the eternal mind of God objectifying nature, is better understood as a normative branch of natural history, that is, as immanent to and inseparable from the natural domain of geophysical systems, and thus an emergent phenomenon of certain forms of the earth's organization of itself, regardless of whether these would be semantically termed "biological" or "technological."[86] From Bateson's point of view, McCulloch had effectively "pulled epistemology down out of the realms of abstract philosophy into the much more simple realm of natural history"[87] by demonstrating that what we experience as the self—the separation between that which is internal to us as subjects in contrast to that which resides externally in the object—is a process of constant renegotiation as a result of our learning, adapting, and growing. A self, in other words, does not stand in an infallible relationship to itself as a pregiven being, as if divinely created according to some eternal blueprint. As such, Bateson's ecological philosophy can be understood as an attempt to naturalize mind without thereby also rendering it mechanistic—or, in his own words, to conceive of mind in a manner that

85 In this context, Bateson was notably influenced by a paper that McCulloch had coauthored with, among others, Humberto Maturana. See Jerome Y. Lettvin et al., "What the Frog's Eye Tells the Frog's Brain," *Proceedings of the Institute of Radio Engineers* 47 (1959): 1940–51.

86 Gregory Bateson, *Mind and Nature: A Necessary Unity* (New York: E.P. Dutton, 1979), 212.

87 Gregory Bateson, "This Normative Natural History Called Epistemology," in *A Sacred Unity: Further Steps to an Ecology of Mind*, ed. Rodney E. Donaldson (New York: Harper, 1991), 216.

could be deemed "neither mechanical nor supernatural."[88] Quite to the contrary, it must continually reproduce itself in response to its environment, and in this ever-ongoing process it must do so by collecting, sorting, selecting, and subsequently decoding information so as to produce differences or provisional limits to its self.[89] Although processes of adaptation, operating in accordance with informational feedback loops, serve to regulate a flow of information, both within the various parts of the observer itself and between the observer and its environment, the fact that the observer is constantly in the midst of becoming rather than self-presently being indicates that we are faced with a holistic matrix within which no single part of the system can exercise unilateral control over the dynamic interaction that constitutes its whole. There is no single kernel within which the essence of a being resides, no control room that individually chooses and executes its actions. Every system is itself a subsystem insofar as it is always defined in relation to an environment, which, according to Bateson, means that we ought to reject the cognitivist presupposition that mind resides solely within the boundary of the animal brain or even within the boundary of the living:

> There is no requirement of a clear boundary, like a surrounding envelope of skin or membrane, and you can recognize that this definition [of mind] includes only some of the characteristics of what we call "life." As a result it applies to a much wider range of those complex phenomena called "systems," including systems consisting of multiple organisms or systems in which some of the parts are living and some are not, or even to systems in which there are no living parts.[90]

88 Gregory Bateson, "Neither Mechanical Nor Supernatural," in Gregory Bateson and Mary C. Bateson, *Angels Fear: Towards an Epistemology of the Sacred* (New York: Bantham Books, 1988), 50.

89 Gregory Bateson, "Form, Substance, and Difference," in *Steps to an Ecology of Mind: Collected Essays in Anthropology, Psychiatry, Evolution, and Epistemology* (London: Northvale, 1987), 455–71.

90 Bateson and Bateson, *Angels Fear*, 19.

In Bateson's terminology, mind is certainly immanent to the circuits of the brain, but the brain itself, in turn, is immanent to the circuits constituting the system he calls "brain-plus-body," and additionally the human body is immanent to the system "person-plus-environment."[91] In the widest sense of the word then, "mind" is an aggregate of parts whose individuation does not preexist their interaction,[92] which parts themselves are not, by nature, in any sense "mental" — insofar as the word "mental" recalls the Cartesian substance dualism that ontologically posits such phenomena as "nonphysical" — but may rather be "geological," "chemical," "biological," and even "technological," even though, importantly, "the objects do not then become a thinking subsystem in the larger mind"[93] but merely part of the circuitry upon which mind as such is conditioned. It is important to note, therefore, that "mind" is not to be mistaken for "consciousness." To call Gaia "mindful," as Thompson does, is not to claim that it is conscious, but precisely the opposite, that the emergence of conscious spirit is conditioned by an expression of unconscious nature. Such is the mereological relationship between part and whole: even though the whole of the mind is complicit in the purposiveness of conscious systems, it can never be fully reported and accounted for within its conscious part.[94]

Hence, as Bateson famously acknowledged, whenever I say "I," it implies, yet covers up, a whole system of spatially and temporally extended relations. Or, as reformulated by Dorion Sagan:

> "my" knuckle-wrinkled, vein-fed fingers stabilized by calcium-containing minerals first showed their mettle in swimming vertebrate ancestors hundreds of millions of years ago. The same chalk that might write these words in white-on-black rather than ink-on-recycled-paper represents calcium

---

91 Bateson, *Steps to an Ecology of Mind*, 317.
92 Bateson, *Mind and Nature*, 92–94.
93 Ibid., 94.
94 Bateson, *Steps to an Ecology of Mind*, 439–40.

> deposits, an ecological waste excreted on the outside of primeval marine cells that would have perished if the toxic calcium ions were not continuously pumped through cell membranes back into the ocean. Moreover, when I sit at my computer, it is not just me, but the mineral precipitates of the earth: silicon, the most common element in earth's crust, along with rare earth minerals collected from multiple continents and requiring such a global distribution and exploitation of labor that it is increasingly the case that no single country, let alone individual, can produce or mass-produce the high-tech products we more or less take for granted.[95]

To begin with, there are apparent traces, in Bateson's dispersion of the modern subject into its terrestrial environment, of the Vernadskian idea of a geometabolic acceleration in the global turnover of materials and energy, not to mention the Teilhardian notion of a planetary attractor that draws history toward increasingly complex patterns of organization, progressively refining planetary consciousness through the construction of artificial platforms to support it. At the same time, recognition of the ostensibly systemic nature of technology in its global impact on our planet dates back at least to the influence that the steam engine had begun to have on the cultural imagination of the West as early as around the turn of the eighteenth century, depicting the agency of technological evolution in terms of natural selection and the struggle for existence in the form of human exercise of self-interest — recall Adam Smith's concept of the invisible hand. Humans, pursuing a competitive advantage, invent new machines, which then eliminate or exterminate the older and inferior ones. In the process, machines "evolve," threatening not so much to become humankind's superior as to environmentally shape its very being. Already in the nine-

95 Dorion Sagan, "Möbius Trip: The Technosphere and Our Science Fiction Reality," *Technosphere Magazine*, November 15, 2016, https://www.anthropocene-curriculum.org/contribution/mobius-trip-the-technosphere-and-our-science-fiction-reality/.

teenth century, Samuel Butler, novelist and critic of Darwin's theory of evolution, observed that humans are so hopelessly committed to their machines that, even if they wished, they could no longer live without them. Less than half a decade after the publication of Hutton's geotheory, humanity had become so dependent upon this mechanical habitat of industrial production, transport, and communication that such artificial systems could be imagined to constitute no less than its ecological niche. Although no single nation is capable of manufacturing a particle collider on its own, perhaps no longer even a color television, it is nevertheless so that, together, as a global organism, we produce a plethora of technological instruments and scientific apparatuses that, though dependent on global flows of energy and resources, and manufactured on multiple continents, now belong to each and every one of us like the vital parts of our own bodies.[96] Following this intellectual genealogy, then, the supposed emergence of the technosphere looks less like a phase transition in natural history than, to quote Julian Huxley, like "new wine in old bottles,"[97] that is, the unconscious return of an old idea in new garb. The natural history of mind evoked by Bateson is an intellectual historical reverberation of a late eighteenth-century Romantic project, whose echoes we can still discern, to this day, in the Anthropocenic sublimation of human artifice into that of a geological force.

§

Despite being a diverse group of scholars, a commonality among the later cyberneticists can nevertheless be found in their general agreement that behind the goal-directed behavior of self-organizational systems lay a much more fundamental and unconscious natural teleology. It was in this spirit that Förster opened the first conference on cybernetic self-organi-

96 Ibid.

97 Julian Huxley, *Essays of a Biologist* (New York: Alfred A. Knopf, 1923), 235–302.

zation with the provocative claim that "there are no such things as self-organizing systems!"[98] Systems achieve their apparent self-organization only by virtue of their interaction with other systems, that is, always within an environment. In a similar vein, Robert Rosen noted the "logical paradox implicit in the notion of a self-reproducing automaton,"[99] and in 1962, Ashby reiterated the main point that "the appearance of being 'self-organizing' can be given only by the machine *S* being coupled to another machine [α ... since o]nly in this partial and strictly qualified sense can we understand that a system is 'self-organizing' without being self-contradictory." Indeed, "since no system can correctly be said to be self-organizing, and since use of the phrase 'selforganizing' tends to perpetuate a fundamentally confused and inconsistent way of looking at the subject," Ashby concluded that "the phrase is probably better allowed to die out."[100] In hindsight, Förster introduced the term "second-order cybernetics" to distinguish their efforts from the concerns of early cybernetics, to mark their preoccupation with the relational and networked nature of communication, and to cement an alliance with the work of Maturana and Varela. Thirty years later, Gordon Pask described the differences between first- and second-order cybernetics by reference to a shift in emphasis from information to coupling, from transmission of data to conversation, from stability to organizational closure, and from external to participant observation.[101] In short, the shift went from Wiener's "command, control, and communication"—using technology,

98 Heinz von Förster, "On Self-Organizing Systems and Their Environments," in *Understanding Understanding: Essays on Cybernetics and Cognition* (Berlin: Springer, 2002), 1.

99 Robert Rosen, "On a Logical Paradox Implicit in the Notion of a Self-Reproducing Automata," *Bulletin of Mathematical Biophysics* 21 (1959): 387–94.

100 William R. Ashby, "Principles of the Self-Organizing System," in *Principles of Self-Organization: Transactions of the University of Illinois Symposium of Self-Organization, 8 and 9 June, 1961*, eds. Heinz von Förster and George W. Zopf (New York: Pergamon Press, 1962), 267–69.

101 Gordon Pask, "Introduction: Different Kinds of Cybernetics," in *New Perspectives on Cybernetics: Self-Organization, Autonomy, and Connection-*

in the instrumentalist sense of the word, to control nature—to a concern that was more analogous to Maturana and Varela's concept of autopoiesis, whereby the artificer itself is intrinsically bound up with the environment that it seeks to govern, thus undermining the Cartesian subjectivity still so central to Wiener's conception of the emancipative possibilities of technology in terms of human mastery. Such is the irony that a movement which in its first-order version sought to naturalize mind did instead end up affirming natural teleology. Arguably, cybernetics never succeeded in resolving the tension—indeed, the contradiction—between these two perspectives: master, control, and design, on the one hand, and complexity, emergence, and self-organization, on the other. More specifically, it never managed to give a satisfactory answer to the problem involved in designing and controlling self-organizing systems, since the realization of such an inhuman biopower of humankind over itself, in its very success, retroactively disclosed the instrumental notion of design and control as an illusion.[102] This is a contradiction inherent to cybernetics as a discipline, which, as noted by Hannah Arendt, we find at the heart of the anxieties of modern humans: whereas the power of humankind to alter its environment increasingly goes on under the stimulus of technological progress, humanity finds itself less and less in a position to control the consequences of its actions.[103]

As if to confirm Arendt's worst fears, Luhmann positively declared, thirty years later, that with the recursive logic of second-order cybernetics, the modern distinction between subject and object had finally been ungrounded in the fundamental self-reference of observing systems—an autopoietic system, as part of its self-making operation, must actively produce and main-

ism, ed. Gertrudis Van de Vijver (Dordrecht: Kluwer, 1992), 24–25. See also Keller, "Organisms, Machines, and Thunderstorms," 71–72.

102 Dupuy, "Cybernetics Is an Antihumanism," 144.

103 Hannah Arendt, *The Human Condition* (Chicago: University of Chicago Press, 1958), 231.

tain its own boundary.[104] Thus, in a higher-order reiteration of the same cybernetic logic, we cannot look at Gaia as a planetary whole without looking self-referentially at ourselves as a part of Gaia looking at Gaia.[105] Philosopher Evan Thompson points out: "In one of his articles Lovelock uses the term *ecopoiesis* to describe Gaia. This term seems just right for conveying both the resemblance and difference between Gaia and the autopoietic cell. The resemblance is due to the ecosphere and the cell being autonomous systems, the difference to the scale and manner in which their autonomy takes form."[106] To bring autopoiesis up to the level of Gaia, then, is to bind Margulis's biological microcosm to Lovelock's geophysiological macrocosm in a positively fractal way, treating all of nature's products as organized wholes, differing not in kind but only in the matter of their degree, that is, isomorphic structures and operations that recur at different scales. Meditating on Margulis's presentations of the bacteria spirochetes in symbiotic association with the eukaryotic protist *Mixotricha paradoxa,* William Irwin Thompson set this vision down as an imbricated form of multiscalar recursion: "So we have a nested universe: the spirochete is in the protist, the protist is in the termite, the termite is in the log, the log is in the forest, the rain forest is in Gaia, and Gaia is inside the solar system, and on and on it goes."[107] If the smallest known autopoietic unit is the bacterial cell, then the largest known is Gaia, for both display the distinguishing features of an autopoietic system, in the sense that as their environment changes, they seek to maintain the structural integrity of their internal organization

104 Niklas Luhmann, "The Cognitive Program of Constructivism and a Reality That Remains Unknown," in *Theories of Distinction: Redescribing the Descriptions of Modernity,* ed. William Rasch, trans. Joseph O'Neil et al. (Stanford: Stanford University Press, 2002), 128–53.

105 Bruce Clarke, "Autopoiesis and the Planet," in *Impasses of the Post-Global: Theory in the Era of Climate Change,* ed. Henry Sussman, 2 vols. (Ann Arbor: Open Humanities Press, 2012), 2:59.

106 Evan Thompson, *Mind in Life: Biology, Phenomenology, and the Sciences of Mind* (Cambridge: Harvard University Press, 2007), 122.

107 William I. Thompson, *Coming into Being: Artifacts and Texts in the Evolution of Consciousness* (New York: St. Martin's Griffin, 1998), 30.

by remaking and interchanging their parts.[108] In either case, the representational notion of objectivity—of a subject looking at an object, representing said object within the boundaries of its a priori conditions of sensibility—is replaced by enactive participation in nature.

In this manner, systems theory eventually seeks to attend to both the elements and the processes of the systems it observes, precisely insofar as in self-referential systems those elements are themselves conceptualized as the product of the processes in which they partake—just as in the Gaian views of Lovelock and Margulis, the form of an organism coevolves along with the form of its environment. Nature is no passive and static object that the subject may control, and the earth is not a cornucopia for the realization of its desires. Such a dualistic conception of a primordial natural condition versus an externally imposed alteration is precisely what Lovelock's and Margulis's challenge to the dichotomy between nature and artifice sought to undermine. As the feminist science studies scholar Donna Haraway has underscored:

> The whole earth, a cybernetic organism, a cyborg, was not some freakish contraption of welded flesh and metal, worthy of a bad television program with a short run. As Lovelock realized, the cybernetic Gaia is, rather, what the earth looks like from the only vantage point from which she could be seen—from the outside, from above[. …] The people who built the semiotic and physical technology to see Gaia *became* the global species, in which they recognized themselves, through the concrete practices by which they built their knowledge.[109]

108 Lynn Margulis, "Big Trouble in Biology: Physiological Autopoiesis versus Mechanistic Neo-Darwinism," in *Slanted Truths: Essays on Gaia, Symbiosis, and Evolution*, eds. Lynn Margulis and Dorion Sagan (Berlin: Springer, 1997), 267, 269.

109 Donna J. Haraway, "Cyborgs and Symbionts: Living Together in the New World Order," in *The Cyborg Handbook*, eds. Chris H. Gray, Heidi J. Figueroa-Sarriera, and Steven Mentor (London: Routledge, 1995), xiv.

On earth, the terrestrial environment has been produced and maintained by organisms as much as these organisms themselves have been transformed and sustained by their environment: the biota actively creates and preserves the conditions for their own subsistence. Technical alteration, in this sense, turns out to be the natural condition for the possibility of life rather than its state of exception, but then again, not in order to impose subjective control. The teleology they refer to is most certainly not one that resides in the subject, but rather, following the Romantic ontologization of Huttonian geology, and subsequently Vernadsky's and Teilhard's posthumanist interpretation of such a self-organizing earth, a natural teleology, a directedness that emerges immanently out of nature's own dynamic processes through precisely the sort of causality that puzzled Kant, whereby the parts do not merely serve to construe the whole, but the whole, in turn, asserts itself top-down upon its parts. What we are confronted with in modernity, then, is not as simple as the loss of unaffected nature — as if, once upon a time, we used to dwell on a planet whose life-sustaining processes remained insulated from the whims of its biota, and is no longer the case, as technology has become a global force on par with that of nature itself — but rather the realization that the biota's active alteration of the environment, including the technological systems of human civilization, was always a fundamental feature of nature's own, inherent productivity. Put in Romantic terms, we are not so much becoming a part of Gaia as we are becoming aware of ourselves as having been part of Gaia all along. "Because we are alive," Lovelock contends, "in a rudimentary way the [earth] system has, through us, become sentient."[110] Nature has always been artificial, which of course means that artifice has always been natural.

Following Arendt, this is the dialectical turn of the first-order cybernetic desire for full control into the second-order erasure of the human that signifies the conclusion of the modern epoch

110 James E. Lovelock, *A Rough Ride to the Future* (Harmondsworth: Penguin, 2015), 24.

and the concomitant subjugation of the earth by planetary technicity. Although Arendt, like her mentor Heidegger, was certainly "worried about the eclipse of the grown by the made,"[111] as the intellectual historian Benjamin Lazier has cogently pointed out,

> some of her most potent language [...] was reserved for the inverted fear: the reduction of the made to the grown. At a still-proximate reserve from the surface of the planet, for example, artifacts and the work required to produce them would appear as those of ants appear to human beings. Our cities would appear as hives, the act of making as the unconscious, unwilled activity of a species[. ...] Arendt may have opposed the eclipse of the grown by the made for fear of doing away with one dimension of the background condition — the biological — out of which human beings emerge. But from a certain remove, that very process appeared as just the opposite: the eclipse of worldliness by earthliness, and the subsumption of human being into the metabolic sway of life and death. The perverse effect of modern technological acumen was to reduce the most artefactual of creatures to mere organisms.[112]

We are fated to an artificial earth, from Arendt's point of view, precisely when we fail to perceive technology at all, or, put differently, when the earth appears to disclose itself all by itself, such that technology is subsumed in nature's self-organization.[113] As has been suggested so far, the Gaian view, in its Lovelockean and Margulian variety, is characterized by such a purported self-presence — an expression par excellence of planetary technicity. Here, the mediatory aspect of technology — the epistemological concern of mediation that characterized the modern epoch — is

111 Benjamin Lazier, "Earthrise; or, the Globalization of the World Picture," *American Historical Review* 116, no. 3 (2011): 613.
112 Ibid., 613–14.
113 Hannah Arendt, "Man's Conquest of Space," *American Scholar* 32, no. 4 (1963): 540.

eliminated by an ontologization of the technological as a vital impulse immanent to nature itself. One may suspect that if Heidegger had been asked to comment on the Gaia hypothesis, he would have discerned therein a characteristically Nietzschean will to power.

Geoengineering as Planetary Medication

It is clear that Gaia, from quite early on in its intellectual historical development, was conceived of as more than a mere hypothesis or research program to be carried out within the modern scientific hegemony of the mechanistic worldview. As pointed out by Latour, in Lovelock and Margulis's work on Gaia "the notions of homeostasis and climate control take on a highly metaphysical dimension."[114] Or even in the words of Lovelock himself, writing in 1979: "[Gaia] is an alternative to that pessimistic view which sees nature as a primitive force to be subdued and conquered. It is also an alternative to that equally depressing picture of our planet as a demented spaceship, forever travelling, driverless and purposeless, around an inner circle of the sun."[115] To properly address Gaia is to acknowledge that, even if only implicitly, it entails a philosophy of nature. Contrary to the modern understanding of the natural world as mechanical and inert, inherited from a tradition that stretches from Galileo through to Descartes, and something to be conquered by humankind — an idea that was revived during the time Lovelock and Margulis were writing in the 1970s in terms of the popular metaphor of "spaceship earth,"[116] which portrayed the planet as a vessel to be rationally guided by a *kybernētēs* — Gaia marks the inheritance of another tradition, with its roots in the Romantic reaction against the instrumental rationalism and scientific materialism that stems from the birth of modern physics, and with its basis

114 Latour, *Facing Gaia,* 123.
115 Lovelock, *Gaia,* 11.
116 Sabine Höhler, *Spaceship Earth in the Environmental Age, 1960–1990* (London: Routledge, 2015), 27–50.

in organicism rather than mechanism. For although circular functions may be instrumental for the self-regulation of certain systems, no system may, according to governing assumptions of the mechanistic worldview, be wholly recursive. This is the reason why the organism came to pose such a challenge to the natural sciences of the late eighteenth century, and what led Kant to posit their self-organization as a regulative principle, that is, as a sort of judicial reservation in an otherwise lawfully mechanistic natural world, a clause only necessary, according to Kant, because of the limitations on human understanding, and not a symptom of some deeper problem in the ontological assumptions of classical mechanics, as the Romantics later came to suspect. Like the circular operation of the governor of a steam engine, as in Hutton's time, or the thermostat of an electric oven, which is the corresponding metaphor we find in Lovelock's writings, the feedback mechanism may be a machine coupled onto a larger system. As with any mechanical contrivance, an oven is thus conceptualized as heteronomous: the input to and outcome of the internal control is determined outside the system by an external operator. The temperature of an operating oven is first determined by the operator and only then maintained within range of that set point by the thermostat. For Lovelock, then, the ban on full recursion is not to be found in the essence of what makes a machine artificial, but in the mechanistic ontology that governs how we think about the operation of entities—with the sole, regulative exception being organisms—in the world, including machines.

A decade after Lovelock's first major publication on Gaia, he restated the concept together with Margulis by pressing the analysis of recursive processes beyond mechanical and computational control processes and toward formal autonomy instead. "Cybernetic systems are 'steered'; biological cybernetic systems are steered from the inside," they declared, and making their affinities with the latter explicit, they consequently went on to clarify that "the Gaia hypothesis postulates a planet with the biota actively engaged in environmental regulation and control *on its own behalf.*"[117] In 1975, when the summer edition of Stewart Brand's *CoEvolution Quarterly* offered the first ever publication on the Gaia hypothesis to a nonspecialist audience, Lovelock and Margulis chose to introduce the reader to the topic by way of a discussion of the seventeenth-century anatomist William Harvey's work on the circulation of blood throughout the human body, which they presented as an analogy, in a similar manner to what Hutton had done almost two hundred years earlier, for the function of the atmosphere as a circulatory system in relation to the totality of the "planetary body."[118] "I speak as a planetary physician," Lovelock would later declare, "whose patient, the living Earth, complains of fever."[119] But Gaia has had a temperature several times throughout natural history, he concluded, and these warmings have occurred well before modern industrialism began to exploit the concentrated energy reserves of fossil fuels.

Notably, Lovelock and Margulis insisted that Gaia demands an ontological upheaval in terms of the fundamental nature of the geological economy of the earth, one that looks to the biological as opposed to the technological domain in order to understand the role of humankind in altering the global environment, that is, by turning to the poietic attitude characteristic of the theory of evolution rather than the instrumental attitude behind the industrial machines that fueled the capitalist economies of the nineteenth and twentieth centuries. Whereas nineteenth-century natural science depicted the evolution of species as an exception to the laws of nature on the pretext that the generation of diversity appeared to contradict the second

117 Lynn Margulis and James E. Lovelock, "Gaia and Geognosy," in *Global Ecology: Towards a Science of the Biosphere,* eds. Mitchell B. Rambler, Lynn Margulis, and René Fester (San Diego: Academic Press, 1989), 9–11 (my emphasis).

118 James E. Lovelock and Lynn Margulis, "The Atmosphere as Circulatory System of the Biosphere: The Gaia Hypothesis," *CoEvolution Quarterly* 6 (1975): 127–43.

119 James E. Lovelock, *The Revenge of Gaia: Earth's Climate Crisis and the Fate of Humanity* (New York: Basic Books, 2007), 2.

law of thermodynamics, Lovelock and Margulis sought to invert the order of priority between the physical and the life sciences in such a manner that the complexification of life would attain the status of natural law. From such a point of view, life, far from being a regulative exception to the natural order, would appear as the supreme expression of nature's self-organization. The point, here, is that the organic interpretation of productivity is fundamentally different from that of its mechanistic counterpart: whereas the ontology underlying the idea of a mechanical contraption renders production conditioned upon the thermodynamic law of depletion and diminishing returns, the idea of a natural organ, on the other hand, eliminates any "natural limit" to production by establishing self-organization and increasing complexity as fundamental to nature in itself. Whereas in the mechanistic sense, industrial production is subject to a finite reserve of resources, the kind of production that is characteristic of life, on the other hand, ought rather to be understood in terms of contemporary debt production, as a process of continuous autopoiesis, a self-engendering of life from life, or a production without conceivable beginning or end. From the perspective of Gaia, the time arrow of life comes to represent a general principle of complexification, running counter to the Malthusian thesis of limits to growth.[120]

From this starting point, Lovelock and Margulis arguably developed an understanding of production profoundly at odds with conservation as an environmentalist value. In his response to the demand for greater regulation of environmental pollution, Lovelock emphasized, quite to the contrary, how the production of waste is an inescapable by-product of what life does naturally: "Pollution is not, as we are often told, a product of moral turpitude. It is an inevitable consequence of life at work. The second law of thermodynamics clearly states that the low entropy and intricate, dynamic organization of a living system can only function through the excretion of low-grade products and low-grade

120 Melinda Cooper, "Life, Autopoiesis, Debt: Inventing the Bioeconomy," *Distinktion: Scandinavian Journal of Social Theory* 8, no. 1 (2007): 33–35.

energy to the environment."[121] Affirming the same thesis, Margulis held that global pollution is so inevitable, so fundamental to the manner in which nature produces itself, that the history of microbial evolution should be understood as a succession of phase transitions to the terrestrial environment, many of them much greater than the contemporary threat posed by industrial capitalism.[122] Humans are far from the first organisms to modify their environment on a global scale so as to cause mass extinction, and their alteration of the atmosphere is orders of magnitude less remarkable than that of other lifeforms, such as, for instance, cyanobacteria. Although certainly fatal to particular organisms, the accumulation of waste, by its very nature, does not threaten life itself. Interpreted as a productive vitality inherent to nature in general, the capacity of life to self-organize is precisely an affirmation of its unconditioned production of itself by itself, such that the innovative aspect of evolution consists in its unrestrained overcoming of what, from the perspective of its finite products, might look like a limit. Nature produces its own provisional limits in order to then expand beyond them, insofar as it can find new ways of organizing itself only through evolutionary solutions that overcome the restrictions that defined the original problem. Of course, entropy remains valid for closed systems, but the boundaries that enclose any given system are precisely what is always under negotiation. Beyond the inevitable depletion of possibilities prescribed by the second law of thermodynamics, the apparently wasteful consumption of matter and energy may, for open systems, give rise to increasingly organized and complex structures. It is the ability of life to render concentrated pockets of energy "for use" that, from the Gaian point of view, constitutes its reason for existing in the

121 Lovelock, *Gaia,* 25.

122 Lynn Margulis and Dorion Sagan, *Microcosmos: Four Billion Years of Evolution from Our Microbial Ancestors* (Berkeley: University of California Press, 1997), 99–114.

first place, since such an intensification of dissipation would be favored in a thermodynamic universe.[123]

Indeed, with its gradient-finding and gradient-reducing capacity, life measurably taps into and spreads available energy even more efficiently than is the case when their organized cycling systems are absent—evolution, as already observed by Vernadsky, is marked by an accelerated pace to the earth's biogeochemical cycling. Encompassing the causes of everything from the earliest of bacteria to our current global technological civilization, such entropy-accelerating systems are in effect depicted as nothing less than an expression of the natural tendency for energy to spread. No different from each other in terms of quality, the emergence of life on earth, around 4 billion years ago, is but a manifestation of the same fundamental propensity in nature to accelerate the dissipation of free energy that lies at the heart of the growth paradigm inherent to our modern capitalist economies. It is merely through an intensification of the same tendency toward the reduction of ambient gradients, by artificially freeing exergy, such as hydrocarbons and radioactive elements trapped in places no other organism has so far managed to reach, that the modern industry of humankind has brought it to new heights. Technological evolution, Margulis contends, "whether [carried along by the] human, bower bird, or nitrogen-fixing bacterium, becomes the extension of the second law to open systems,"[124] such that the propagation of entropy-producing artificial systems, and the environmental ramifications in their wake, is an inevitable outcome of nature simply doing what it does.

In the writings of Lovelock, such ontological assumptions are curiously combined with a loathing of environmental regulation, which culminated in his defense of nuclear energy as the

---

123 Dorion Sagan and Jessica H. Whiteside, "Gradient-Reduction Theory: Thermodynamics and the Purpose of Life," in *Scientists Debate Gaia: The Next Century,* eds. Stephen H. Schneider et al. (Cambridge: MIT Press, 2004), 173–86.

124 Margulis and Sagan, *Acquiring Genomes,* 47.

solution to the imminent depletion of fossil fuels,[125] and his support for large-scale technological intervention into the earth's geospheres.[126] But it is not so much the implied ecomodernism that is remarkable — it is, after all, an already well-established tradition whose own internal consistency is as complicated as the relationship between first- and second-order cybernetics, and that therefore can be said to include a number of diverse figures from Richard Buckminster Fuller to Brand[127] — as that it issues forth out of an attitude to technology that can best be described as "vitalist": it is because technological organization is negentropic that it seeks to grow beyond any provisional closure, and insofar as such systems are self-organizing, we ought to dismiss all efforts to artificially limit them. Succinctly summed up by the political economist Melinda Cooper, "this is a vitalism that comes dangerously close to equating the evolution of life with that of capital,"[128] rediscovering bourgeois relations of production in nature, as Friedrich Engels famously remarked on Darwin's theory of evolution. Just like life, technology is perceived as inherently progressive, expansionist, and evolutionary, possessing the same vital impulse to perpetuate itself for no other sake than the perpetuation of the same process, namely, the reproduction of itself as an end in itself. Made into the metaphysical foundation of an inherently dynamic nature, the vital impulse toward increasingly complex organization ontologically proceeds from the organic process of becoming, unrestricted by the closure of any sort of limitation or fixed form, and since it knows no limits, it inevitably ends up embracing all possible domains, living or nonliving, biological or technological, natural or artificial.

---

125 James E. Lovelock, *The Vanishing Face of Gaia: A Final Warning* (New York: Basic Books, 2010), 105–18.
126 Lovelock, *Revenge of Gaia,* 194–95.
127 Fred Turner, *From Counterculture to Cyberculture: Stewart Brand, the Whole Earth Network, and the Rise of Digital Utopianism* (Chicago: University of Chicago Press, 2006).
128 Melina Cooper, *Life as Surplus: Biotechnology and Capitalism in the Neoliberal Era* (Seattle: University of Washington Press, 2008), 42.

Such a perspective, Cooper suggests, may be appropriately labeled an "affirmative vitalism," since it seeks to stage the values of process and becoming against the apparently reductive principles of erstwhile mechanism — be it in terms of "being," "closure," "disciplinarity," or "stasis."[129] On the other hand, it is precisely in the lack of any principled limits that there is a tendency for the becoming of such a generalized vitalism to collapse everything into itself, thereby also extending self-organization from the organism to nature as a whole. Hence, if "vitalism," in the strict sense of the word, entails some substance or force that would serve to demarcate animate and inanimate matter, then Lovelock's and Margulis's conception of technology cannot, admittedly, be said to be vitalist. However, if by "vitalism" we mean the manner in which life is made fundamentally material, and the behavior of inanimate matter, in turn, is made lifelike in its capacity to self-organize, then we can begin to discern the sense in which vitalism plays a crucial role for the kind of artificial earth implied by the Gaia hypothesis. It is not merely that, throughout natural history, no species is immutable, but rather, no organized system capable of entropy production whatsoever is immutable. With Huttonian deep time, and later evolutionary theory, pervasive change replaces eternal fixity in more than one way, because for Lovelock and Margulis, evolution encompasses not only the struggle for survival among organisms, but also, more generally, a natural-teleological striving toward ever-greater complexity and creativity in nature as such.

Perhaps the affirmative vitalism at the heart of Lovelock and Margulis's Gaian discourse was best described in the mid-1990s by Brand's associate editor of the *Whole Earth Catalogue*, Kevin Kelly, who, in the appropriately titled *Out of Control* (1994), chronicled various examples of how artificial systems had been "discovered" by systems theorists to be ontologically equivalent to living systems. Indeed, the organic structure of nature, Kelly asserted, is increasingly expressed in the function and behavior

129 See also Thomas Osborne, "Vitalism as Pathos," *Biosemiotics* 9 (2016): 185–205.

of our economies, industrial processes of production, and information networks, which, as they become more sophisticated and complex, gradually acquire the same adaptive, self-organizing, and decentralized features that many million years of evolution had already given rise to in the organism. As a self-proclaimed prophet of the proliferation of hybrids, Kelly preaches the therapeutic dissolution of this confoundment in the antinomic conclusion that humankind is undecidably situated both inside and outside of nature at once. As he sees it, the cybernetic marriage of nature and artifice demonstrates the twofold meaning of technology — "supplement" versus "supplant" — to be a false dichotomy insofar as it fails to grasp the self-productive becoming that marks the absolute identity underlying both. Describing this most natural impetus toward gradient-reduction as "a network of vital life, an outpouring of a nearly mechanic force that seeks only to enlarge itself, and that pushes its disequilibrium into all matter, erupting in creatures and machines alike," Kelly goes on to depict such metabiotic forms of autopoiesis as a "circle of becoming, an autocatalytic set, inflaming itself with its own sparks, breeding upon itself more life and more wildness and more 'becomingness,' [...] [with] no conditions, no moments that are not instantly becoming something more than life itself."[130] So stereotypically Nietzschean is Kelly's proposed definition of the vital impetus in nature to express itself in evermore complex and creative forms that his effort to assure the reader that his "description of the aggressive character of life is not meant to be a postmodern vitalism"[131] appears at best a vacuous claim without substance, if not a misunderstanding of the terminological complexity of both "postmodern" and "vitalism." In any case, according to Kelly, equilibrium is death, precisely as Lovelock predicted,[132] which means that it is when we are con-

130 Kevin Kelly, *Out of Control: The New Biology of Machines, Social Systems, and the Economic World* (Reading: Perseus Press, 1994), 98.

131 Ibid., 97.

132 Ibid., 83.

fronted with perpetual dynamism that we know we are faced with lifelike processes.

But it is crucial to note the preestablished harmony that underlies the surface upon which the particular forms of nature are perpetually created and destroyed. For here, the very will to power has ontologically migrated so as to constitute an inherent feature of nature in itself. As in Huttonian geology, there is thus no room for a genuinely catastrophic scenario, such as the extinction of life, for death is but an epiphenomenon, that is, a transgression of temporary forms of life in the name of an increased vitality overall. As in Vernadskian geochemistry, evolution is a slow process, and it might very well be faced with setbacks and delays, but at the end of the day, it is an evolution nonetheless. From this perspective, the main trends of evolution — increases in number of organisms, species, and taxa, increases in number of cell types, and, despite periodic setbacks, increases in biodiversity and its phenomenological correlates, increases in information processing, and increases in social communication — are all regarded as expressions of a fundamental vital impetus toward increased entropy production, whereby complexity and energy expenditure go hand in hand. As the next step in natural evolution, technology constitutes the medium wherein ever-more energy can be stored and deployed. In the words of Margulis and Sagan, for humans "to bed down with electrical artifacts and electronic fabrications is, in short, entirely natural — entirely in keeping with life's ancient tendencies to expand, pollute, and complexify. It is our second nature and the nature of all of our ancestors."[133] If the revolution caused by Darwin revolved around his depiction of humankind as neither apart from nature nor exceptional in relation to it, but as the product of a broader evolutionary process, then his intellectual legacy found its own evolutionary continuation in Teilhard de Chardin's vitalist account of the genesis of technol-

133 Lynn Margulis and Dorion Sagan, "Welcome to the Machine," in *Dazzle Gradually: Reflections on the Nature of Nature* (White River Junction: Chelsea Green Publishing, 2007), 77.

ogy as an indispensable part of life, which, in turn, was taken up by Lovelock and Margulis's notion of an earth that is thoroughly artificial — by its very nature — insofar as it is always in the process of being produced by the biota.

Seen from the Gaian perspective of Lovelock and Margulis, then, technology is far more ancient than the naturalistic-anthropological fetishization of humankind as toolmaker suggests, and much more fundamental than the exclusively instrumental status ascribed to it by the mechanistic worldview would let on. Indeed, the fabrication of hard minerals by primordial bacteria — as an organ for their continued flourishing, produced in the process of co-opting their environment — was in full swing long before animals or even plants had appeared. More to the point, the claim is not merely that fire-making flints, bone tools, and stone fishing weirs coevolved with the human species long before documented history, nor even that primates and other animals develop techniques and utilize tools too, but that lifeforms perpetually alter the conditions for their own existence. For Margulis as for Vernadsky, life is "less a thing, and more a happening, a process,"[134] thereby dismantling the ontotaxonomical distinction between living organisms and their nonliving environment in favor of a flat plane of immanent vital tendencies in nature to spontaneously organize itself. And since life is processual, it is impossible in principle to neatly delimit where the organism ends and its environment begins. As opposed to an anomaly in a natural world characterized by nothing but dead mechanism, there is a ubiquity of life in the work of Margulis and Vernadsky: nature is teeming with liveliness, and it erupts in all manner of different couplings between physical systems that together organize matter into progressively more complex forms. Margulis writes affirmatively of how "technology, from a Vernadskian perspective, is very much part of nature[. ...] The plastics and metals incorporated in industry belong to an ancient process of life co-opting new material

134 Margulis and Sagan, *What Is Life?*, 50.

for a surface geological flow that becomes ever more rapid."[135] Such flows are certainly not modern, nor premodern, nor even human. Because ever since it evolved from its bacterial origins, the biota has relentlessly produced, shaped, and transported the earth's rocks, soil, water, and air. No matter their particulars, all lifeforms require energy and matter, and in getting and spending this energy and matter, life alters its environment in species-specific ways.

Fundamental to life, then, is that it is evolving, expressing itself in the will to transgress the boundaries that define its natural-historically contingent manifestations. What starts out as pollution from a growing population may become raw materials as a species mutates or instigate the right conditions for another species to thrive. From bacteria to shrubs, from marine worms to insects, lifeforms reroute and reuse their waste. Since previously inaccessible or unexploited niches constitute new opportunities for life to further expand, evolution favors the selection of traits that creatively open up new "markets" for the exploitation of resources hitherto deemed void of use value. Echoing the Schumpeterian idea of "creative destruction," which describes how capitalism's process of industrial mutation incessantly revolutionizes the economic structure from within, as innovation destroys long-standing arrangements and frees up resources to be deployed elsewhere,[136] the inherent tendency of life, like that of capitalism, is to incorporate more and more of its environment into itself in order to expand and intensify its production. Long before the appearance of Homo sapiens upon the face of the earth, an increasing number of chemical compounds was enrolled and utilized by the proliferation of different lifeforms. Hence, prehuman organs, such as barium sulfate spines and calcium shells, were early expressions of this tendency, and human artifice, in turn, merely extends this trend in nature.

---

135 Ibid., 51–52.

136 Joseph Schumpeter, *Capitalism, Socialism, and Democracy* (London: Routledge, 1994), 81–86.

Whereas new technology may disconcert us, most artifacts in our lives are, as Heidegger pointed out, so familiar to us — so "naturalized" — that their being used as tools is noticed only when they break down and cease to serve the function that they used to.[137] The point for Margulis, of course, is that the very sense of artificiality is context-dependent. It is hardly a surprise, from Margulis's point of view, that Heidegger's concern in the middle of the twentieth century with technologies such as hydroelectric dams already seems to us, if not a case of Luddism, then at least severely old-fashioned. The boundaries of "the natural" have since symbolically moved. Our present causes of concern are rather with robotics and genetic engineering, or, to extend upon the theme of global enframing that worried Heidegger, we hardly raise an eyebrow toward the practice of controlling rivers through damming as we are instead inching ever-closer toward considering how to alter the entire planet's climate through geoengineering. However, the question of the "artificiality" of technology is not restricted to the symbolic domain, because it obviously has material consequences too. Infrastructures for accessing clean water and sanitation, modern medicine, or even the institution of money as a medium for universal commodity exchange, not to mention such basic techniques as the lighting of fire, the sewing of clothes, or the preservation of food, all constitute the habituated background of our everyday environment, and they do so to the point that the average human today, if it finds itself in an environment deprived of these techniques and technologies, certainly does not thrive, if it manages to survive at all. Ought not these technologies, therefore, also be considered crucial parts of our natural environment, in the literal sense of that word? Because, for both Margulis and Lovelock, the natural has to be produced. Although this production of nature, as Margulis has argued, does not necessarily need to be viewed from the selfish scenario of the survival of the fittest, but is better understood in symbiotic terms, it is nevertheless a

137 Martin Heidegger, *Being and Time*, trans. Joan Stambaugh, ed. Dennis J. Schmidt (New York: SUNY Press, 2010), 72–75.

case of active alteration through extended technical means. If the organism comes into being in an environment more or less suited to its existence, the boundaries that determine its flourishing are by no means fixed but are constantly negotiated by the organism's own actions.

§

Echoing Teilhard de Chardin, Lovelock proceeds even further to state that technological evolution has now eclipsed evolution by natural selection as described by Darwin, and he retraces the historical origin of this astonishing transition back to the invention of the artifact that dominated the industrial scene during the time Hutton was writing, namely, Thomas Newcomen's steam engine, which was first drafted in its original model in 1712. In fact, the coal-powered engine's capacity to perform sustained work exceeding one kilowatt is Lovelock's preferred thermodynamic definition to mark the onset of the Anthropocene, as the ensuing positive feedback loops would propel an exponential growth in material flows, human populations, and information processing. According to Lovelock, the year 1712 thus signifies a turning point in the history of the earth, when humankind gained access to cheap and abundant energy as it learned to "make use" of fossil fuels, and, thereby, for the first time in history, linked heat with work.[138] Until the Newcomen steam engine, humans had received heat energy from wood, and work energy from wind, water, and oxen, but the two had seldom met — except instantaneously, in the barrel of a gun. In the wake of the industrial revolution, however, nearly all work starts out as heat — now transformed into the barrel of a cylinder. Since the beginning of the eighteenth century, increasingly sophisticated engines have enabled the growth in complexity to gain steam at a breakneck pace and have allowed humans to transform not only their local milieu but the entire terrestrial environment.

---

138 Lovelock, *A Rough Ride to the Future*, 14–16.

In this sense, Gaia, as a cybernetically enhanced organism, is like a macrocosmic counterpart to the project envisioned by Manfred Clynes and Nathan Kline, who jointly coined the term "cyborg" in 1960 to refer to the kind of self-modifications on part of the organism that they thought would facilitate the next stage in the human vocation, namely, the colonization of outer space.[139] In an eminently Teilhardian manner, Clynes and Kline opened their paper by declaring that "space travel challenges mankind not only technologically but also spiritually, in that it invites man to take an active part in his own biological evolution."[140] Whereas in the past, natural evolution would favor the altering of bodily functions to better suit the organism's exploitation of various environments, now, with humans beginning to grasp the homeostatic functioning of their own bodies, they argued that it would be possible to bypass heredity altogether through physiological, biochemical, and electronic modifications to humankind's existing *modus vivendi*. In other words, adaptation would become active and self-directed rather than passive and indirect. On the one hand, this new technologically guided evolutionary process is paradigmatically different because it marks the end of the primacy of evolution by natural selection, which has carried us and the earth for the past 3 billion years at its relatively slow, unhurried pace. In comparison to that of natural selection, the rate of technological change marks a new stage in evolution precisely because of its potency in radically altering the terrestrial environment in a comparatively short amount of time. Yet, on the other hand, technological innovation as the main driver of evolution is merely the

139 It seems quite likely that Lovelock, during his sojourns in the world of NASA exobiology, would have attended to Clynes and Kline's essay in the 1960s. In fact, Lovelock has cited this source in his most recent work, but it seems as if he has misconstrued their argument in favor of outfitting the human body with technical prostheses for the notion of a self-organizing machine that animated Wiener's concern with cybernetics. See James E. Lovelock, *Novacene: The Coming Age of Hyperintelligence* (Cambridge: MIT Press, 2019), 29.

140 Manfred E. Clynes and Nathan S. Kline, "Cyborgs and Space," in *The Cyborg Handbook*, eds. Gray, Figueroa-Sarriera, and Mentor, 29.

most recent manifestation in a long line that mirrors nature's organization of itself. It is this inherent dynamism of nature and its capacity to produce novelty that was crucial to the making of a planet that could support modern humans, which remains critical to our reshaping of the earth in the Anthropocene, and which will continue to be decisive for our future survival. The driving force of technological—just like zoological—evolution is the utilization of energy to convert explosive force into rotary motion: to alter materials from one form to another, and in doing so to produce a surplus of resources and a concomitant proliferation of life.

This seems at first glance promising for those modernists who, like Wiener, desire instrumental control over the globe, because, surely, there is nothing that humans understand as well as their own artifice? Although already present in ancient Greek philosophy, such as in Plato's *Republic* (c. 380 BCE), wherein the notion of certainty indirectly rested upon examples drawn from the craftsmanship characteristic of *technē,* the same idea would later prove extremely suitable for the linear causality of mechanistic philosophy, and thus returned with force during the Enlightenment. The most famous adherent among modern philosophers is arguably Thomas Hobbes, for whom the highest certainty is found in purely artificial constructions, namely, entities that humans have put together piece by piece, and therefore can also take apart in the same manner. In *Leviathan* (1651), perhaps his most famous work, Hobbes depicts the modern state as an intricate device—through the metaphor of a well-designed machine, later to be picked up by Hutton—which, in its ideal form, mirrored God's perfection of mechanism in the human. The heart of the human and the social body is compared by Hobbes to that of a spring, nerves to strings, and joints to wheels, all of which, in concert, gave motion to the entire body as previously envisioned by the artificer. According to this machine metaphor, the right kind of motion is ensured by installing just the right kind of components into the contrivance. In principle, at least, such a contrivance could be constructed with absolute perfection and certainty. As the political theorist Langdon Winner has pointed out, it was thus imagined that the artificer — in this case, the political or legal philosopher — would know, in advance, the complete contents of well-established order, because there is no room for emergent surprise if the animation of the body consists entirely of the motion of its elementary constituents.[141] Ideally, there is nothing that the artificer should not be able to anticipate. In order to understand the whole, the case is merely one of calculating the sum of its parts. And, vice versa, if we are confronted by a contrivance that mystifies us, we need only take it apart piece by piece to study its components.

This deterministic *Weltanschauung*, which reigned hegemonic during the eighteenth, nineteenth, and well into the twentieth century — following not only the mathematical work of Isaac Newton but also that of Gottfried W. von Leibniz, Leonhard Euler, and Joseph-Louis Lagrange, and also the philosophical inquiries of René Descartes and Auguste Comte — strongly supported a paradigm of order founded on the principles of reductionism and predictability. It was believed that a given set of causes would lead to the same known effects, implying that the behavior of a system could be explained by the sum of the behavior of its parts. This kind of system would be predictable in the sense that once its global behavior was defined, future events could be anticipated by introducing the correct inputs into the model; its process would flow along orderly and predictable paths where, given a set of initial conditions, one could confidently predict the outcomes. The notion of causal determinism was famously advanced as a thought experiment by Pierre-Simon, Marquis de Laplace — also known as "Laplace's demon" — where he maintained that a being with the knowledge of the precise location and momentum of every atom in the universe could calculate their past and future values for any given time using the laws of classical mechanics:

141 Langdon Winner, *Autonomous Technology: Technics-out-of-Control as a Theme in Political Thought* (Cambridge: MIT Press, 1977), 279–80.

> We ought then to regard the present state of the universe as the effect of its anterior state and as the cause of the one which is to follow. Given for one instant an intelligence which could comprehend all the forces by which nature is animated and the respective situation of the beings who compose it — an intelligence sufficiently vast to submit these data to analysis — it would embrace in the same formula the movements of the greatest bodies of the universe and those of the lightest atom; for it, nothing would be uncertain and the future, as the past, would be present to its eyes.[142]

The universe as described by Laplace is a mechanistic world defined by differential equations, operating much in the same manner as a mechanistic contraption. To know something thoroughly, one would know it in the same sense that an artificer knows its own work and its products — or, as for Leibniz, the way in which God knows his clockwork universe. Indeed, the artificer would know it right down to its very soul, the source of its life and motion, and if there is anything like a telos to such a machine, it is the intention according to which it was created in the first place. For nineteenth-century physicists, then, the demon was a technologically sophisticated being — a pointsman on a railway, a strategist sitting at his telegraph wires, in short, a *kybernētēs* — an intelligent being capable of controlling and directing individual elements from the ideal position of a properly all-encompassing perspective, standing outside of the universe looking in. Thus the artisanal nature of the demon was important: it combined the skills of architect and builder, ontologically grounding the idea of the existence of a complete plan for the entire universe.

But as the attentive reader might have already predicted, the understanding of the essence of technology implicit in the Gaia hypothesis is decidedly different. If we like to think that at least the artificer understands its own invention, not even this is the

142 Pierre-Simon Laplace, *A Philosophical Essay on Probabilities*, trans. Frederick W. Truscott and Frederick L. Emory (New York: Wiley & Sons, 1951), 4.

case, argues Lovelock. In fact, not only does the invention of an artifact rarely take place without more than a tiny proportion of specialists having the slightest idea of how it was made or how it works, but moreover, only rarely is there enough foresight on the part of the inventor to even begin to grasp its future potential to be altered, developed upon, and adopted for different purposes.[143] And so, once an artifact has been constructed, the potential for unintended consequences and runaway scenarios is simply too great for the possibility of control. Not at all dissimilar to the Gaian discourse, Winner, in 1977, at around the same time that Lovelock and Margulis were developing their hypothesis in response to its first reception, described this conception in terms of autonomous technology run amok: "In its very nature, this is not a soul that is planned in advance or inserted into the machine by design. Instead, it is a quality of life and activity springing from the whole after the myriad of parts have been fashioned and linked together. A ghost appears in the network. Unanticipated aspects of technological structure endow the creation with an unanticipated telos."[144] We construct artifice without understanding how it works, and by the time we do realize some of its implications, many of the changes that it has brought about are already irreversible. Technology is beyond our control, if by "us" we mean an autonomous subject in the Cartesian sense of the term. Arguably, Lovelock would agree with Hobbes that artifice does have a willful, active, and self-determining quality of its own. But this kind of agency is not restricted by an external artificer, or, as in the case of social contract theory, an omniscient lawgiver. Instead, what Lovelock voices is the immanent autonomy of technology generated within the interlocking dynamic of interchangeable parts, built in bits and pieces over hundreds or thousands of years of technological evolution, but without anything like divine providence or a preordained control mechanism to oversee and steer the interaction of these parts from without:

143 Lovelock, *A Rough Ride to the Future*, 62–64.
144 Winner, *Autonomous Technology*, 280.

> How much of [the Anthropocene] can be attributed to the inspiration of talented individuals and their flowering during and after the Renaissance? More likely, I think, the burgeoning progress we see around us, good and bad, may have come from a simpler and cruder source: that is, the work of rude mechanicals who worked blindly like Wagner's Nibelungen, who made their Ring with no thought for the consequences.[145]

This passage expresses a key theme in Lovelock's reasoning on technology: we severely overestimate the sway of instrumental reason over technological evolution. Indeed, science excels at providing an explanation after the fact, and even then its theories must be constantly revised or discarded. But to imagine that it could stay ahead of its own project, always with intention and in control of where it is going, is for Lovelock an absurd idea.

Writing about his own experiences as an inventor, Lovelock admits that it could sometimes take him years to partially understand or explain inventions that came into his mind almost as instantly as if "a gift from intuition."[146] In a passage that recalls the Romantic attempt to reconcile practical and theoretical philosophy, he describes how invention is as much a response to the call of nature to bring its products forth as it is a case of the artificer imposing its own will upon it:

> While inventing I am in touch with that inner layer of the mind where information is processed without awareness. As a scientist I have to explain each step of a process openly, and often I do this on a piece of paper, with pencilled sums or simple diagrams. In real life these two processes tend to merge, and scientific intuitions come from the inner layer without my being aware until the completed thought emerges.[147]

---

145 Lovelock, *A Rough Ride to the Future*, 22.
146 Ibid., 63.
147 Ibid., 64.

There is no doubt that the Romantics would have approved of this revival of feeling and instinct, which conjures up the ideal of the intuitive participation in nature that they celebrated as characteristic of the artist. This groundlessness is unthinkable precisely because of its primacy to thought, preceding the formation of the subject, and it is precisely when gazing vertiginously into the abyss that the fundamental creativity of nature shines through the superficial appearance of objective finitude proposed by the mechanistic worldview. Attended to in earnest, nature reveals itself to be infinite and sublime in its productive power. It is only because the modern subject perceives itself as being capable of manufacturing for us a finite and ordered world that this infinity is covered over and obscured from our view. Clearly, Lovelock's emphasis on the limitations of the instrumental rational attitude, and the importance he ascribes to intuition as perhaps being the only way of developing a feeling for a system of nature that fundamentally resists formalization, has Gaia in mind. The becoming of Gaia can only be schematized as a static model, which means that such a formalization simply cannot account for the unconditional groundlessness that is its perpetual production of itself by itself.

The presence of this kind of inversion of ancient Greek geosomatism in Gaian discourse has only been made more palpable in light of Lovelock's last suggestion — in *Novacene* (2019), his own writings on the Anthropocene — that we require an explicitly "cyborg" solution to the threat of global environmental change. In a manner remarkably similar to Teilhard de Chardin, Lovelock begins therein by recovering humanity's spiritual value. It is humanity's glory as the supreme progeny of a unique planet to reflect the universe back to itself.[148] Yet, Lovelock's hymn to the human is subdued — precisely as in Teilhard de Chardin's celebration of the impending posthuman — by the twist in the argument telegraphed by the book's subtitle: *The Coming Age of Hyperintelligence*. Simultaneously, then, the Novacene — which is Lovelock's own preferred terminological substitute to the

148 Lovelock, *Novacene*, 23.

Anthropocene—will be the epoch, arriving as we speak, in which biological life, in the form of human beings, yields its status as apex cogitator and passes the evolutionary reins to electronic life. Consigned by Gaia to an exclusively functional role, humans have no choice but to submit to their artificial systems, albeit in line with what can be best described as a badly masked theodicy, assuring that their surrender of control is ultimately in line with nature's tendency toward increasingly complex and creative expressions. According to Lovelock's narrative, life, by the end of the twentieth century, found ways to jump the chasm between the natural and the artificial, the born and the made, the biological and the technological. The world is quickly being populated by living, complex, cyborgic systems destined to evolve, diversify, merge, diverge, and transform continually toward some fundamentally unknowable, organic future. In the natural history of our planet, as retold by Lovelock, artifice thus comes to appear as a kind of Teilhardian phase transition in natural evolution, across which the will to power ventures in order to produce a new world. No longer at the center even of civilization, humans, even as they bear an obligation to facilitate the delivery of its noöspheric stage, are thereby released from responsibility for what the earth will become. For if humanity is the controlling artificer of a new world, then it bears a terrible responsibility for gambling with the future. But if, on the other hand, the vital force—in Nietzschean terms, "the will"—of evolving life itself is ultimately responsible for enlivening their technologies, then humans are suddenly off the hook.[149]

These days, of course, there is no shortage of cultural references to the existential risk of bootstrapping artificial superintelligence. But there is plenty of evidence to suggest that Lovelock's late turn toward a heightened interest in machine intelligence was merely making explicit an aspect that was implicitly present in Gaian discourse all along. For although the idea of an impending machine takeover admittedly seems rather foreign to the Gaia hypothesis as such, it fundamentally draws on the

149 Bryant, "Whole System, Whole Earth," 250–58.

same natural evolutionary tendency toward increasingly complex forms of self-organization through nature's infinite productivity. It is informing to note, for instance, that the silicon beings of the impending Novacene already resonate with Margulis's development, in the late 1980s, of an autopoietic Gaia in which the machinic components extruded by an advanced, technological civilization are reincorporated into the geological economy. First outlined in "Gaia and the Evolution of Machines" (1987), a think piece published in the *Whole Earth Review,* itself a descendant of Brand's *Whole Earth Catalogue,* Margulis and Sagan complemented their autopoietic description of Gaian operations by explicitly adding global technology to the layer of the earth's concentric spheres: "Not only are members of the more than 10 million existing species components of the Gaian regulatory system, but so are our machines."[150] Following in the vein of Butler's Lamarckian challenge against his own contemporary Darwinians, they go on to insist that "although not by themselves alive, like viruses and beehives, machines are capable of reproduction, mutation and evolution,"[151] in effect transferring the ontogenetic qualities misplaced by the discipline of biology as the prime criteria of the individual organism into the metabiotic interface of living matter, geophysical formations, and technological organizations that continually remix the earth system in hybrid forms. This is to suggest that, over and above the propagation of the biota, biological life and machine reproduction are, in our age of the Anthropocene, now interdependent:

> The reproduction of technological societies and their components is part of the autopoiesis of the biosphere[. . . .] From a biospheric view, machines are one of DNA's latest strategies for autopoiesis and expansion. The classification of machines as non-autopoietic and nonliving does not negate the fact

150 Dorion Sagan and Lynn Margulis, "Gaia and the Evolution of Machines," *Whole Earth Review* 55 (1987): 15.

151 Ibid.

> that they reproduce, and reproduce with mutation, as avidly as viruses. Like beehives, termite mounds, coral reefs, and other products of the activity of life, machines — if indirectly through DNA and RNA — make more of themselves. Through us they make other machines.[152]

Hence also the celebration of technical production in Gaian discourse: the ritual of pointing to a nature profligate beyond expectation in order to petition that no assertion, no matter how arrogated or unreasonable, is not somehow adequate to nature's potency to surprise in its blind profligacy. Here, creative destruction is conscripted as the exception that always disproves the rule, disabusing us of the constraints of ever suffering the imposition of having to select the correct. In this, the antique idea that no part of nature can be inconsolably illegitimate, because all legitimacies are never not eventually realized, inverts into the conviction that no intentional state can be illegitimate because nature has the power to actuate anything and everything through the mindless maximalities of its myriad becomings.

Constructive Chaos

Although it certainly includes aspects of both technological hubris and environmentalist sentiment, the intellectual history of the Gaia hypothesis is not exclusively a narrative about the instrumental impetus expressed in postwar cybernetic control discourse, the Cold War, and the space race, let alone a feel-good tale about humanity's recovery — often associated with Earthrise and an accompanying shift in our cultural self-consciousness — of a premodern intimacy with nature that had supposedly been lost with the rise of mechanistic philosophy. Put in the context of a genealogy of planetary technicity, it has been suggested that the intellectual history of Gaia is better understood dialectically, as a story of *mereological confusion, historical*

152 Ibid., 19.

blindness, and *ontological forgetting,* insofar these three capture precisely how the instrumental and the poietic operate in and through one another. First, of confusion: to look at our planet as a blue marble is in fact to gaze upon the fully self-present and integrated earth-organism as a pictorial artifact, the all-encompassing earth system as a world picture. Second, of blindness: to focus on how the view of the earth from space was overtly mobilized is to miss some of the more subtle effects of this sight after we cease to register its novelty, after we cease, in a fashion, to see it, namely, the very mode of disclosure whose interpretative horizon is closed down as a result of its holistic pretense. Third, Gaia tells a story of forgetting: many have espied in Gaia a neopagan icon of nascent global environmental awareness inspired by the one-body metaphor of a naturally grown superorganism, but if we follow the lead of Heidegger and Arendt, we must ask whether Gaia has always been an artificially enframed globe in disguise, and whether the supposedly all-inclusive notion of a global environment is really the worst kind of enframing that reduces the radical otherness of nature to qualitatively uniform use value.[153]

The implication of this dialectic is perhaps best borne out in McLuhan's assessment, at first glance puzzling, that the constitution of the earth as an object of technical decree was enabled — not contested — by the Romantic conception of the planet as a work of art, that is, something poetically crafted from within itself rather than instrumentally manipulated from without. Such a McLuhanesque verdict would stress that global systems-thinking is characterized by an ambivalence whereby the technological figuration of life and of nature is inseparable from the naturalization and animation of technology. This double movement, which is already characteristic of the parallel conception of machines and human bodies working in the industrial age, is nevertheless heightened in the information age, where computer programming, modeling, and simulation become dominant techniques of knowledge and of biopolitics,

153 Lazier, "Earthrise," 626–27.

effectively subsuming the entirety of earth under the banner of "world dynamics."[154] But, as we have seen, the further narrowing of the gap between nature and artifice that is mirrored in earth science during the latter half of the twentieth century is not as one-sided as the mere mechanization of life, or, in other words, not as one-sided as the reduction of the organism to the machine. Even paradigmatically control-oriented disciplines, such as cybernetics, born out of a desire to engineer and calibrate nature in much the same manner as one calibrates a machine, ended up eventually acknowledging that a quest for full control would have to turn self-reflexively toward itself.

As the science studies scholar Donna Haraway noted in her foreword to *The Cyborg Handbook* (1995), although "Gaia — the blue- and green-hued, whole, living, self-sustaining, adaptive, auto-poietic earth — and the Terminators — the jelled-metal, shape-shifting, cyber-enhanced warriors fighting in the stripped terrain landscapes and extraterrestrial vacuums of a terrible future — seem as first glance to belong in incompatible universes,"[155] it is crucial, if one is to properly grasp the intellectual nuance of the cybernetic resources enrolled by Lovelock and Margulis, to probe the surface of such modern philosophical prejudices in order to excavate therein the ontological tensions that complicate any reductive illustration of the hypothesis as either a Promethean or a neopagan symbol. Indeed, Haraway's exploration of Gaian discourse sought to record precisely how "Lovelock's earth — itself a cyborg, a complex auto-poietic system that terminally blurred the boundaries among the geological, the organic, and the technological — was the natural habitat, and the launching pad, of other cyborgs."[156] Whether Lovelock ever read Haraway's *Cyborg Manifesto* (1985) is doubtful, but the latter's concept of the cyborg was anyway but a mobile figure for the various real and imaginary incursions of informatic instrumentalities into organic systems, intended to flesh

154 Neubert and Wiemer, "Rewriting the Matrix of Life," 11–12.
155 Haraway, "Cyborgs and Symbionts," xi.
156 Ibid., xiii.

out the cultural and intellectual environment of the aforementioned space race and the Cold War military-industrial complex around which it coalesced — a witty but attentive excursion into that quasi-academic territory within which Lovelock himself was working. To be sure, Haraway's interpretation of Gaia as an ideological figuration of systems theory remains particularly astute in the deconstruction of its prevalent caricature as, in the sarcastic retort provided by Margulis herself, "an Earth goddess for a cuddly, furry human environment."[157] For it certainly seems odd that chiefly responsible for a hypothesis so named was an inventor and NASA contractor, as opposed to, say, "a vegetarian feminist mystic suspicious of the cold war's military-industrial complex and its patriarchal technology."[158] As we have seen, however, and as is hinted at by Haraway, it is no coincidence that the somatic metaphors of planetary ailment — "the earth has a fever" — were closely coupled with the technical language of medical diagnosis and cure, because the intellectual inversion of the earth as an organic body into that of an artificially modified cyborg has its roots precisely in the modern desire for a technologically ordered globe.

In fact, Kirchner already noted as early as the late 1980s that the "two groups that immediately embraced Gaia were environmentalists and, paradoxically, industrialists. The former argued that harming any part of the planetary 'organism' could have far-reaching consequences, while the latter argued that Gaia's capacity for homeostasis made pollution control unnecessary."[159] For what, from the Gaian point of view, are the planetary health risks — to use Lovelock's physiological metaphor — of geoengineering? "Nothing we do is likely to sterilize the Earth,"[160] Lovelock proclaims, but the consequences of global technological intervention could hugely affect the conditions of survival for certain carbon-based lifeforms, such as humans. If we are

157 Margulis, "Gaia Is a Tough Bitch," 140.
158 Haraway, "Cyborgs and Symbionts," xiii.
159 Kirchner, "The Gaia Hypothesis," 224.
160 Lovelock, *Vanishing Face of Gaia*, 155.

to believe Lovelock, one thing is certain, though: Gaia will find other ways of self-regulating. It might be, he speculates, that electronic lifeforms based on semiconducting compounds will evolve to take over that task, but at a completely different habitable state, perhaps many degrees warmer than is suitable for humankind. But it is critical, from the Gaian point of view, that we do not normatively condemn the imperative to constantly expand production since it merely expresses the vital tendency in nature to exploit previously untapped resources in order to increase complexity. Global environmental change is not the fault of a mode of production that is premised upon the lack of limits, but rather, in the words of Lovelock, "the constructive chaos that always attends the installation of a new infrastructure."[161] This does not mean that we ought to strive to use technology, in the radically diminished instrumental rational sense of the word, in an effort to return to some supposedly harmonious and balanced, "natural" state. In fact, such is the modern fallacy of the fetishization of the Holocene. Rather, the lesson of Gaia is that the notion of an ideal condition — whether it be in terms of a utopia toward which humans ought to strive or an Edenic state from which humankind has since fallen — must be discarded in favor of living together with, or in the midst of, the complex and feedback-oriented economy of nature. Instead of imagining that we, as planetary stewards, are responsible for correcting what is out of our control, we must attune ourselves to its self-correction. For instance, humans are already concentrating themselves in increasingly complex clusters, so rather than hubristically ascribing themselves the task of saving the entire earth, or of restoring some artificial version of a supposedly "natural" climate, why not live a comfortable life in air-conditioned megacities instead? Again with a reference to the realm of biology, this serves bees, ants, and wasps perfectly well, Lovelock points out, as well as, incidentally, he notes, the inhabitants of Singapore, who come closest to Lovelock's ideal model of future

---

161 Lovelock, *A Rough Ride to the Future,* 12.

living.[162] Such is the response to global environmental change that would be poietically in line with nature's self-organization, namely, to fully embrace humankind's artificial alteration of its environment as precisely its natural condition. Not according to some preestablished blueprint, such as through a return to the Holocene, but rather in line with the open-ended imperative to transgress every "natural" limit or boundary that would place humans under the illusion that they dwell within a closed system.

162 Ibid., 121–23.

CONCLUSION

The Will to Terraformation

In "Was the Anthropocene Anticipated?" (2015), Clive Hamilton and Jacques Grinevald argue that the idea of humanity as a geological force has its intellectual roots in the birth of the discipline of earth systems science in the 1980s, and thus emerged on the academic scene without much of anything that can be said to resemble a rich genealogical lineage. Their claim is quite straightforward: the "scientists in the nineteenth and first half of the twentieth centuries did not possess the modern scientific concept of the Earth system of which the Anthropocene is an outcome."[1] It is because of the earth science of the late twentieth century, they insist, that we are now able to perceive the danger of global environmental change for what it really is, namely, an anthropogenic affair, and one that may spell the end of humanity as a species. But as our preceding genealogical investigation in this book has indicated, such a purported artificialization of the earth was not simply abstracted from empirical evidence. The Anthropocene is a phenomenon whose pedigree is to a nontrivial degree philosophically implicated. Rather than the product of an objectifying gaze, as if observing the earth from afar, the question concerning the ontological status of collec-

1 Clive Hamilton and Jacques Grinevald, "Was the Anthropocene Anticipated?," *Anthropocene Review* 2, no. 1 (2015): 60.

tive humanity as a geological force, as it has figured in the context of the modern study of the earth, has grown out of a concomitant concern with human subjectivity. When we are faced with anthropogenic global environmental change, such a phenomenon cannot be reduced to the empirical observation that humanity is altering the global environment. On the contrary, it also involves a concern with the status of this being—the human.

Elsewhere, Hamilton has identified a "theodicy-like structure"[2] behind the idea that humankind merely actualizes nature's striving toward complexity—what he calls an "anthropodicy," a secularized version of the theological defense of the ultimate benevolence of the whole and of the goodness that in the end transcends and defeats contradiction, but where the same belief in the goodness of human-directed progress has come to replace that of God.[3] In light of our genealogical investigation here, the structure of theodicy proposed by Hamilton captures very well what is essential to planetary technicity: in spite of environmental devastation and mass extinction of species, it stands as a silent reassurance that all this death and destruction is but an inevitable by-product of nature's inherent tendency toward the creative destructive production of increasingly complex patterns of organization.[4] In the earth system paradigm, nature is depicted as being already caught in antagonistic interaction with collective human labor. But this is not because we never encounter nature in itself—as if with

2 Clive Hamilton, "The Theodicy of the 'Good Anthropocene,'" *Environmental Humanities* 7, no. 1 (2015): 236.

3 Ibid., 234, 236.

4 In fact, even those earth scientists who defend the centrality of deep time for humanity's self-awareness of itself as a driver of global environmental change recurrently concede that "geological timescales are often used as reasons for non-action on societal, intra- and intergenerational timescales ('climate has always changed,' 'coral reefs have become extinct several times, but reappeared' and so on)"; Jan A. Zalasiewicz et al., "Petrifying Earth Processes: The Stratigraphic Imprint of Key Earth System Parameters in the Anthropocene," *Theory, Culture & Society* 34, nos. 2–3 (2017): 98.

the disappearance of the human, nature would return to perfect harmony. On the contrary, it is held that the fiction of a stable nature, only secondarily disturbed by human intervention, is wrong even as an inaccessible ideal that we may approach if we withdraw as much as possible from our activity. Hence if nature is seen as in itself already disturbed, out of joint, far from equilibrium, and creatively destructive, then human labor is simply an expression of what we might tentatively call a "will to terraformation," one understood to be inherent to nature, such that humankind's drastic modification of its terrestrial environment seemingly remains in perfect harmony with the inherent discord of the earth system. Downscaling the technological alteration of nature would thus constitute the most artificial response to our global environmental predicament.

Even though it is certainly not false, then, as the earth scientists Charles Langmuir and Wallace Broecker pointed out in the mid-1980s, that without an understanding of the earth as a homeostatic system, dynamically circulating around interlocking feedback loops, the scientific concern for global environmental change could not have arisen,[5] this is not the same as to conclude that the birth of earth system science also marks the inauguration of the kind of nonlinear causality that positions humankind as an important *part of*—rather than *apart from*—the terrestrial environment. For as they stressed in the preface to the same book: "If there is one theme that we hope comes through [as crucial for the idea of humanity as a geophysical force], it is of a connected universe in which human beings are an outgrowth and an integral part."[6] And as we have seen in the preceding genealogy, this theme is central to the concern for humankind's being on the earth at large. Rather unsurprisingly, Langmuir and Broecker's illustration of a terrestrialized planet by way of an internalization—and thereby also naturaliza-

5 Charles H. Langmuir and Wallace S. Broecker, *How to Build a Habitable Planet: The Story of the Earth from the Big Bang to Humankind* (Princeton: Princeton University Press, 1985), 16–17.

6 Ibid., xv.

tion — of technological production is close not only to the geophysicist Peter Haff's proposition that the global technological systems of the Anthropocene mark the emergence of an additional geosphere — "the technosphere" — but also to Vernadsky's and Teilhard de Chardin's insistence that modernity brings about a natural evolutionary phase transition to the structure of the terrestrial environment, what they called "the noösphere." In both cases, the protagonist of global environmental change is the earth itself, not humans. Indeed, insofar as the preceding genealogy has persuasively traced what has herein been called "planetary technicity," it should by now have become clear that the *longue durée* history of the Anthropocene is best understood as a history of the ontological flattening of being and the concomitant death of the modern subject.[7] Any sense of a limit, such as an interpretative horizon belonging to human history, thereby disappears into the depth of geological time, and the subject suddenly vanishes from the center of the global environmental drama. Ironically so, since the purported novelty of the Anthropocene is precisely its anthropogenic dimension.

Such is the lingering theodicy of the Anthropocene: the corruptibility of the synthetic merge between natural geomorphology and human artifice in the Anthropocene thus consists in the risk that the affirmation of the earth's complete artificialization thereby also eliminates any real sense of a threat to nature. Instead, anthropogenic manipulation becomes but an expression of nature merely doing what nature does — it produces, and endlessly so, without condition, and in accordance with no other end than its maximal self-expression. As such, we have pointed in the same direction as the historian Dipesh Chakrabarty, who holds that the efforts "by Earth system scientists to communicate the message of the current planetary environmental cri-

7 Let me make this extra clear. The history of the Anthropocene, I claim, is, among other things, and certainly not exclusively so, a history of the death of the modern subject. Nevertheless, I claim that a historiography of the Anthropocene that does not take into consideration this crucial dimension fails to grasp what it is that we inherit when we labor with it conceptually in the present — hence my disagreement with Hamilton and Grinevald.

sis" — Langmuir and Broecker's included — "speak necessarily in two voices, [... a] human-centered and [a] planet-centered [voice]."[8] Contra Charkrabarty though, these two voices do not correspond to a humanistic and a natural scientific perspective, respectively. Rather, both are complementary components of the same scientistic enframing of the earth that sees instrumentalism and organicism conjoin in unholy matrimony — the same ambivalence between full freedom and full determinism that we have traced from Hutton to Lovelock, whereby the complete enrollment of the entire planet as standing reserve is accomplished only through the erasure of the human and its concomitant incorporation into a presubjective will to terraformation. In other words, the same theodicy-like structure undergirds the understanding of the essence of technology that figures in the current geoscientific discourse on the Anthropocene — there is precisely a genealogy here.

In fact, it should be telling that Hamilton and Grinevald ultimately fail to grasp the undecidable arrest of the twofold meaning of technology into the understanding that underlies the contemporary concept of the earth system, namely, the notion that, as they themselves even write, "human history liberated from the natural history of the Earth, has been wiped away, because, as [...] Chakrabarty has told us, the two histories have now converged, giving us a kind of hybrid Earth, of nature injected with human will."[9] It is precisely this technological destruction of subjectivity that we have been tracing from Hutton to Lovelock: the death of the modern subject and the reinscription of artifice into natural history. The turn from linear to nonlinear causality is decisive for the reaffirmation of the organic conception of the earth in modernity, whose lineage stretches through Lovelock back to Hutton. It might be reasonable to argue, as Hamilton and Grinevald do, that "the giants of natural history, when thinking about civilized man as a geological force,

8 Dipesh Chakrabarty, "Anthropocene Time," *History & Theory* 57, no. 1 (2018): 25.

9 Hamilton and Grinevald, "Was the Anthropocene Anticipated?," 68.

lived in a world unaware of a disturbed global nitrogen cycle, a mass extinction event, and global climatic change due to the atmosphere's changing chemical composition," but it is, on the other hand, intellectually historically unsubstantiated to draw the conclusion that they therefore were unable to conceive of the kind of nonlinear causality that underlies a proper "Earth system thinking, the science of the whole Earth as a complex system beyond the sum of its parts."[10]

§

But the enframing of the earth does not merely eliminate subjectivity, because without its point of reference, objectivity dissolves too. As I have argued in this book, this is a feature of the earth system, not a bug. On the one hand, its systematic ambitions are oriented toward describing the dynamics of the planet through the identification of causal relationships. Underpinning this approach is the view that an observed phenomenon can be explained in terms of cause and effect, where the ideal endpoint is the abstraction from the observed causal relationship into generalizable laws about the structure of the earth system in order to govern it. The goal in studying the earth holistically is to understand it well enough to explain past changes and predict future events, and to do so as accurately as possible, that is, by modeling the earth systematically, as an organic whole. On the other hand, however, whenever there is a multiplicity of forces acting in conjunction, the outcome, and the explanation, will be contingent on all those contributory processes. In this manner, even if we were to be able to identify and specify the initial conditions of a given process operating in a particular environment, this would not be enough to be able to predict its outcomes with any great degree of certainty. The impasse in deciding the antecedent resides in the fact that, whereas the structure of a system regulates the interrelationships between its components, the

10 Ibid., 61–67.

interrelationships between its components, in turn, regulate the structure of the system.

The atmospheric chemist Sarah Cornell and her colleagues at the Stockholm Resilience Centre have argued that "absolute objectivity — in the sense of the traditional scientific ideal of single-variable experimentation under controlled conditions — is not possible in Earth system research, so other ways are needed of ensuring adequate scientific scrutiny and verification of claims."[11] There are serious methodological and conceptual challenges to understanding the dynamics of the earth, especially in terms of how models can capture the feedback between its components.[12] In fact, the concept of the earth system issues a caution: we ought to always remain mindful of the circumstance that a system as intricate and interlinked as earth will not respond like a mechanical clockwork. Two inconsistent insinuations can thus be draw from this perspective, as demonstrated in the State of the Planet Declaration:

> In one lifetime our increasingly interconnected and interdependent economic, social, cultural and political systems have come to place pressures on the environment that may cause fundamental changes in the Earth system and move us beyond safe natural boundaries. But the same interconnectedness provides the potential for solutions: new ideas can form and spread quickly, creating the momentum for the major transformation required for a truly sustainable planet[....] Such systems can confer remarkable stability and facilitate rapid innovation. But they are also susceptible to abrupt and rapid changes and crises.[13]

---

11 Cornell et al., "Earth System Science and Society," 31.

12 Evan D.G. Fraser et al., "Assessing Vulnerability to Climate Change in Dryland Livelihood Systems: Conceptual Challenges and Interdisciplinary Solutions," *Ecology and Society* 16, no. 3 (2011): 1–12.

13 Lidia Brito and Mark Stafford Smith, *State of the Planet Declaration — Planet under Pressure: New Knowledge towards Solutions Conference, London, 26–29th of March 2012* (London: Diversitas, 2012), 5–6.

On the one hand, this indicates a certain faith in the predictability of the interactions that constitute the operation of our planet, so as to allow for the possibility to manipulate the earth system according to will. Yet, it simultaneously involves the recognition of irreducible and emergent phenomena, interlevel causation, and sensitivity to initial conditions, which poses serious challenges to modeling. Cornell and her colleagues again:

> Human society is an intrinsic part of the Earth system while it is also the collective mind observing, researching and intervening in its dynamics. The balance of experimentation, observation, abstraction, theoretical coherence, and subjectivity or objectivity all determines what constitutes scientific evidence. Global environmental change has been a nexus for change in this balance, reframing our role as scientists within the system that we study.[14]

Ironically, the sophistication with which we are able to measure, experiment with, and manipulate the planet has only made us more uncertain about the most basic questions concerning our own ability to know and thus control the observed processes of change. But it is obvious that this is not first and foremost an epistemological problem. As pointed out in the quotation just above, it has to do with the question of humankind's being on the earth.

Of course, Chakrabarty is right to point out that, in the earth system paradigm, the idea that the human can be successfully demarcated and identified in accordance with its artificial undertakings has become untenable in the light of its metaphysical shift from stability to instability. But it was never the intention to put the human under erasure. Rather, the impetus around which global change research emerged in the late twentieth century itself has a genealogy that spans a rich history of predicting the natural limits of our planet; only this time, in our current paradigm, the shift in question has reoriented

14 Cornell et al., "Earth System Science and Society," 30.

our understanding of humankind's relationship to the earth in such a manner that, all of a sudden, the idea of natural limits apparently denies the entire reality of natural history—and the point, here, is that we are not principally dealing with empirical evidence, but with a hermeneutic shift in the interpretation of what the empirics indicate. It is in these terms that quantifiable models of the planet's carrying capacity, such as Thomas Malthus's eighteenth-century calculations of the earth's ability to sustain human populations, and the efforts of the Club of Rome in the 1970s to measure the limits of socioeconomic growth, have in hindsight been argued to have generally done a poor job. As is often argued, the Western tradition, since at least René Descartes, has operated on the basis of a misleading ontological dualism, which has inhibited humanity from comprehending its proper relationship to the planet, that is, as an active participant in the production of an already artificial earth. Humans, just like any other organism, transform their environment to sustain themselves—it is something that we currently do, and most importantly, something that we have always done.[15] As a matter of fact, no earth system scientist would deny that. Rather, these scientists are the first to admit, and even emphasize, that our planet's inhabitability is not down to some predetermined threshold, but to the fact that any suitable environment emerges first from the immanent production of temporary states of stability, and, moreover, that every stable state must be continually reproduced or else it will dissolve into disorder.

However, there is a bitter pill of irony at the heart of such a conclusion: in the Anthropocene, the unnatural power of technology has supposedly grown so great that it has come full circle. In other words, as human artifice is deemed to have become a "natural force," in the literal sense of the word, there is a tendency to circumscribe the account of the human self altogether by conceding to the "scientific fact" that it is nature, today, and not artifice, that is essentially unnatural. Neither subject nor

15 Arnulf Grübler, *Technology and Global Change* (Cambridge: University of Cambridge Press, 1998), 345–47.

object, such a force of nature is conceptually converted into unconditional production by and for itself. Perhaps most forcefully affirmed by the self-proclaimed "vital materialist" Jane Bennett, our distinctively normative status as humans — with its basis in "concepts as the norms that determine what we have made ourselves responsible for, what we have committed ourselves to, and what would entitle us to it, by particular acts of judging and acting"[16] — thus vanishes together with the subject in the same ontological void that subordinates human to natural history. Bennett notes that "in the long and slow time of evolution [...] mineral material appears as the mover and shaker, the active power, and human beings, with their much-lauded capacity for self-directed action, appear as *its* product."[17] Although human history plays out in accordance with this blind drive, humans are never in control, because the underlying force itself is rather akin to that presubjective agency that Friedrich Nietzsche, by drawing upon the resources of Schopenhauerian animism, sought to articulate with his cosmic vision of the world as will to power.

So, what becomes of history when we are invited to let go of the foundational divide — since Wilhelm Dilthey marked out the territory of the human sciences — between historical and deep time? When we as human scientists are encouraged to extend our perception to the past tens of thousands of years, what becomes of narrative and interpretation? Have we reached the expiry date of historicity itself, when the scale of history, consumed in the depth of geological eons, becomes as vast as to defy hermeneutical interrogation altogether?[18] This concern has been astonishingly absent from the Anthropocene debate, but it was first raised by the philosopher Catherine Malabou:

16 Robert B. Brandom, *Articulating Reasons: An Introduction to Inferentialism* (Cambridge: Harvard University Press, 2001), 33.

17 Jane Bennett, *Vibrant Matter: A Political Ecology of Things* (Durham: Duke University Press, 2010), 11.

18 Cécile Roudeau, "The Buried Scales of Deep Time: Beneath the Nation, beyond the Human ... and Back?," *Transatlantica* 1 (2015): 1–2.

> Chakrabarty *denies* any metaphorical understanding of the "geological." If the human has become a geological form, there has to exist somewhere, at a certain level, an isomorphy, or *structural sameness,* between humanity and geology. This isomorphy is what emerges — at least in the form of a question — when consciousness, precisely, gets interrupted by this very fact. Human subjectivity, as geologized, so to speak, is broken into at least two parts, revealing the split between an agent endowed with free will and the capacity to self-reflect and a neutral inorganic power, which paralyzes the energy of the former. Once again, we are not facing the dichotomy between the historical and the biological; we are not dealing with the relationship between man understood as a living being and man understood as a subject. Man *cannot appear* to itself as a geological force, because being a geological force is a *mode of disappearance.*[19]

In order to clarify what Malabou means by a "mode of disappearance," it is crucial to recognize that the disclosure of the earth in terms of an ontologically flat system simultaneously conceals that the deployment of the vocabulary of the natural sciences — including that of earth science — is itself an artificial phenomenon, such that the study of natures, including the endeavor of earth system science, itself has a history, and that its own nature, if any, must be approached through the study of that history. For as human artifice is deemed to have become a "natural force," there is a risk that we are blinded precisely to the historical process by which the norms that articulate the contents of the concepts applied are artificially instituted, determined, and developed.[20] Chakrabarty explains:

---

19 Catherine Malabou, "The Brain of History, or, the Mentality of the Anthropocene," *South Atlantic Quarterly* 116, no. 1 (2017): 40–41 (my emphasis).

20 I use the term "historical," here, to refer to the particular sense that "human history" — as opposed to "natural history" — acquired in the German, *Geisteswissenschaftliches* tradition, alluding in particular to Georg W.F. Hegel's naturalization of the Kantian picture of conceptual norms by taking those norms to be instituted by public, social, recognitive practices,

> The need arises to view the human simultaneously on contradictory registers: as a geophysical force and as a political agent, as a bearer of rights and as author of actions; subject to both the stochastic forces of nature (being itself one such force collectively) and open to the contingency of individual human experience; belonging at once to differently-scaled histories of the planet, of life and species, and of human societies.[21]

However, the folding of humankind back into natural history — rendering its artificial features natural again, precisely by rendering nature itself essentially unnatural — has not first and foremost benefited the principle of caution, nor has it primarily cultivated a sense of responsibility. It is in the face of such an enthusiasm to eagerly jump into the abyss that we may fully sense the gravity of Malabou's concern with, as she puts it, "the mentality" of the Anthropocene. To be sure, it is not the intellectual genealogy per se that is disconcerting, merely the forgetting of our conceptual horizon that occurs when the history of the study of natures is subsumed back into natural history, that is, when the contingency of our socially instituted normative vocabulary is naturalized in such a manner that it dissolves us as concept-users of our responsibility and robs us of the possibility of holding other concept-users accountable for theirs.

---

thereby bringing the noumenal origins of this normativity "back to earth by understanding normative statuses as social statuses — by developing a view according to which [...] all transcendental constitution is social institution" (Brandom, *Articulating Reasons*, 34). This tradition holds that, although of course artificial activities arise within the framework of a natural world, artificial products and activities become explicit as such only by the use of a normative vocabulary that is in principle not reducible to the descriptive vocabulary the natural sciences. Note, however, that the deployment of the vocabulary of the natural sciences is itself an artificial phenomenon, that is, it is possible first with the use of concepts, and is therefore something that becomes intelligible only within a conceptual horizon that is already artificially instituted, which, incidentally, is contingent upon the history of social institution studied by the human sciences.

21 Dipesh Chakrabarty, "Postcolonial Studies and the Challenge of Climate Change," *New Literary History* 43, no. 1 (2012): 14.

§

As they outlined in an oft-cited article, Andreas Malm and Alf Hornborg have offered a similar critique of the Anthropocene that draws upon the Marxist notion of reification, that is, the consolidation of social relations into natural laws. "We find it deeply paradoxical and disturbing," Malm and Hornborg declare, "that the growing acknowledgement of the impact of societal forces on the biosphere should be couched in term of a narrative so completely dominated by natural science."[22] Instead, they argue that the prevailing focus on the human species as the culprit of global environmental destruction results from our ignorance of the technological condition that is in fact its real cause, not because these conditions are overlooked, but because they are made into a "force of nature," such that a certain mode of disclosure appears as the natural properties of things-in-themselves. They are writing, in other words, about the naturalization of technology. The fundamental error in Anthropocenic reveries, Malm and Hornborg argue, is the same one that Marx underlined in his remark that industrial production comes to be "encased in eternal natural laws independent of history, at which opportunity *bourgeois* relations are then quietly smuggled in as the inviolable natural laws on which society is founded."[23] What Malm and Hornborg find disturbing in the Anthropocene discourse is that it champions an ontology that obliterates the dualism between subject and object in the service of nothing but further technical manipulation.[24] As a consequence, they perceive the call for a posthuman subsumption of humankind into its techno-ecological environment to be

22 Andreas Malm and Alf Hornborg, "The Geology of Mankind? A Critique of the Anthropocene Narrative," *Anthropocene Review* 1, no. 1 (2014): 63.

23 Marx, quoted in Malm and Hornborg, "The Geology of Mankind?," 67.

24 Malm and Hornborg, "The Geology of Mankind?," 62. See also Andreas Malm, *The Progress of This Storm: Nature and Society in a Warming World* (London: Verso, 2017).

dogmatically in line with the disclosure of planetary technicity rather than critically questioning it.[25]

Hornborg notes that references to "nature" in modern Western discourse have often been implicitly normative, evoking those aspects of reality that are purportedly self-evident and, as a consequence, unquestionable and unalterable. By making use of "social metaphors that transfer meanings from relations in the human world to relations with the nonhuman one, committing societies to specific trains of thought,"[26] such a transfer forces social critique to surrender before the instrumentalization of nature by the capitalist machinery.[27] The sociologist Luigi Pellizzoni puts it succinctly: "There is an important point implicit in the neoliberal blurring of naturalness and artefacticity[. ...] It is the role played by the transfer of concepts and ideas from one domain to another; a transfer during which they may lose their metaphorical status to gain a descriptive one; a transfer, in other words, that may entail an unrecorded shift from 'as' to 'is.'"[28] It is in this manner that the enframing characteristic of planetary technicity drains terrestrial habitation of any sort of purposeful earthly dwelling by flattening the diversity of semiotic valuations into an ontologically uniform status as standing reserve, such that anything and everything can be translated into qualitatively empty use-value and thus enrolled into

25 Alf Hornborg, "Artifacts Have Social Consequences, Not Agency: Toward a Critical Theory of Global Environmental History," *European Journal of Social Theory* 20, no. 1 (2017): 95–110, and Andreas Malm, "Against Hybridism: Why We Need to Distinguish between Nature and Society, Now More Than Ever," *Historical Materialism* 27, no. 2 (2019): 156–87.

26 Alf Hornborg, *The Power of the Machine: Global Inequities of Economy, Technology, and Environment* (New York: Altamira Press, 2001), 197.

27 Characteristic of the enframing of planetary technicity is that the calculative mode of instrumental reason comes to entirely dominate thinking as such, which is the clue as to why it establishes a nihilistic (meaningless) production without end — a capitalist will to power of endless growth in GDP for the sake of no higher value than production for productions own sake.

28 Luigi Pellizzoni, *Ontological Politics in a Disposable World: The New Mastery of Nature* (London: Routledge, 2015), 67.

the nihilistic project of endless economic growth. Guided by a technological-vitalist mythos that conceives of the manifold of beings on earth as devoid of intrinsic worth and as nothing but resources for the sake of an empty principle of a planetary-scale will to terraformation, the Anthropocenic engine that gradually accelerates the rate at which species after species is made extinct is provided ideological steam precisely by the naturalization of technology, namely, through the misrecognition of the culturally and historically contingent logic of capitalism for the universality and necessity of natural law. It is crucial to remember that in the Anthropocene discourse, it is precisely not nature that is unalterable and unquestionable, but technology. For it is technology, in the wake of modernity, which has taken over those aspects of nature that could still render writers such as Edmund Burke awestruck and humble in the face of the sublimity of geophysical phenomena as late into the modern epoch as only thirty years before Hutton presented his "Theory of the Earth" (1788). It is technology, to reiterate Fredric Jameson's observation, that has come to replace nature as the domain of the sublime in the cultural imagination of the West.[29] Through an ontological sleight of hand, artifice has come to take on life-like properties — mind, will, and spontaneous motion — which places it in rebellion against any kind of valorization of the human, regardless of whether such a valorization be religious or secular, ontotheological or post-Heideggerian.

It is important to note that neither Malm nor Hornborg advocates the return to essentialism understood as substance. On the contrary, they draw attention to the fact that already Marx himself made the point that there is no "original nature" left anywhere. Yet, this did not lead Marx to question the centrality of the human subject. As Malm argues, the human, for Marx, "like any other species in the material world, [...] is forever tied to nature, but the nature of the ties is never natural."[30] It is per-

29 Fredric Jameson, *Postmodernism, or, the Cultural Logic of Late Capitalism* (Durham: Duke University Press, 1991), 34–38.
30 Malm, *The Progress of this Storm*, 162 (my emphasis).

haps in the famous "Fragment on Machines," in the *Grundrisse* (1939), that Marx breaks most explicitly with the instrumentalism of his fellow young Hegelians by raising the concern of "the appropriation of living labor by objectified labor,"[31] that is, the inverted relationship between human and machine whereby the former takes on a prosthetic role in service to the latter. Importantly, Marx does not do so by adopting a poietic attitude in its place, but by critically investigating the resubjectivization of the human insofar as it begins to take on habits and patterns of thought particular to a certain mode of disclosure under capitalism. In the very process by which the natural world is instrumentally taken apart, compartmentalized, rendered equivalent, and made manipulable, Marx notes how, ironically, the instrument itself comes to appear as an autonomous entity "whose unity exists not in the living workers, but rather in the living (active) machinery, which confronts his insignificant doings as a mighty organism."[32] Here, the privilege ascribed to intentional construction, alteration, and improvement is overturned. Instead, technology takes on organic features: no longer dependent on human planning and construction, it grows in unpredictable ways, evolving its structures according to what seems to be its inner self-organizing capacity: "In machinery, objectified labor itself *appears* not only in the form of product [...] but in the form of the force of production itself. The development of the means of labor into machinery is *not an accidental moment* of capital, but is rather the *historical reshaping* of the traditional, inherited means of labor *into a form adequate to capital.*"[33] Seen in the light of Marx's concept of fetishism, then, it is not because of the complex interaction of individuals in modern society that industrial capitalism emerges—as Vernadsky and Teilhard de Chardin, or even Lovelock and Margulis, would have it—as an automatic, self-organizing system of nature's production. Quite

31 Karl Marx, *Grundrisse: Foundations of the Critique of Political Economy*, trans. Martin Nicolaus (Harmondsworth: Penguin, 1973), 693.

32 Ibid.

33 Ibid., 694 (my emphasis).

the contrary, it is because of our belief in the doctrines of complexity and emergence that we are predisposed to act in such a manner that the instrumentalist logic of capitalism can realize its empty project of production for nothing but production's own sake, which can be accomplished only insofar as the world is disclosed in such a manner that humanity ends up treating even itself as but an instrument.

Just as Marx recognized that every culture operates within a certain shared understanding of the human and its place in nature, so it is, today, that we are prisoners to our own particular horizon of understanding. This is a historical argument, not an essentialist one. Most importantly, it recognizes, *pace* theorists of the Anthropocene such as Chakrabarty, that the prominence of antiessentialist dogmas is as historically conditioned as that of essentialist ones. If, in the Anthropocene, we find ourselves overwhelmed by the complexities of the world, and to such a degree that we have come to discard the question of essence altogether, then this is because planetary technicity discloses the world in a manner that presupposes its ontological flattening. And it is precisely in this ontological flattening, this antiessentialist pretense of having done away with the question of the meaning of being, that we find the essence of planetary technicity, and thus the essence of our current understanding of technology. A number of commentators on the Anthropocene condition, Chakrabarty included, have made important contributions by demonstrating how the globalization of technology destabilizes the permanence of the Enlightenment conception of the human as both a moral and an epistemological subject, but we must avoid taking this new dispersed subject at face value, and instead ask what allowed for its appearance. Put simply, the earth system paradigm, insofar as it is premised upon the disclosure of the world in an ontologically indifferent light, would for Marx be an example of ideology par excellence, and the concomitant naturalization of technology a symptom of fetishism. Because an organicism that depicts the production of commodities as the result of self-propelling assembly lines conceals precisely the labor that drives these factories, and thus allows for industrial capitalism to be perceived as the natural embodiment of the flow of abstract exchange value. It is a textbook case of the resurgence of myth in modernity, with the help of which "the bourgeoisie transforms the reality of the world into an image of the world, History into Nature."[34] Through this mythologization, which grants technology an autoproductive and self-determining nature, it symbolically begins to occupy that groundless abyss whose ceaseless activity underpins the modern world. Technology, once it becomes sensible as a properly global phenomenon, causes itself, determines everything without itself being determined, and thus comes to constitute the only end in itself in relation to which everything else comes to take on but a mere instrumental role.[35] The vitalization of the technological, therefore, turns on an understanding that technology is by nature intrinsically expansive: its law of evolution is one of increasing complexity rather than entropic decline, and its inherent creativity is autopoietic rather than adaptive. In the wake of the Anthropocene, it is no longer nature that is imagined to be beyond human categories and capacities. Quite to the contrary, nature has been eclipsed by the artificial. But this is not the same as to say that humans are therefore in control. From the point of view of the earth system paradigm, our technology has on the contrary come to make up an entirely new compartment of the planetary body, to which the purposes of humankind are now subordinated.

This kind of substantivist belief rests upon a fatalistic attitude toward the historical effect of technology. Although such a fatalism need not necessarily be negatively charged, but may rather be a perverse kind of cheerful fatalism, it is nevertheless so that it runs directly counter to any social critical endeavor to preserve a meaningful sense of agency in a world of increasingly asymmetric power relationships, thereby undermining our emancipative capacity to break free from the nihilistic atti-

34 Roland Barthes, *Mythologies*, trans. Annette Lavers (New York: Noonday Press, 1991), 140.

35 Marx, *Grundrisse*, 692.

tude toward technical production as an end in itself.[36] Attempts to conceptualize global technology by granting it its own life-like attributes—such as Vernadsky's and Teilhard de Chardin's noösphere, Lovelock and Margulis's Gaia, or the more recent technosphere—are thus perfectly in line with the task of articulating a new myth in an effort to reinscribe the relentless reorganization of the world by industrial capitalism—with its concomitant destruction of the institutionalization of meaning and purpose—into, again, a meaningful and natural order of things. Only this time, insofar as such a mythologization is successful, artificial production and technological manipulation will no longer appear to us as an external imposition upon an otherwise stable and static nature. Quite to the contrary, they will be depicted as being perfectly in line with the ontologically primitive disorder of the natural world, and in this sense technology comes to be seen an expression of the immanent will to terraformation, that is, as but the execution of that most natural—unquestionable and unalterable—tendency for the earth to self-organize. This is a case of fetishism precisely for the reason that, as Hornborg has argued, technology cannot be sufficiently understood solely in terms of the physical work it performs.[37] Rather, it also embodies "tacit assumptions about [its] rationality and efficiency."[38] Hence, a central aspect to the social function of technology is that it relies on beliefs about its nature. The implementations and operations of technology, rather than mere expressions of the intrinsic and objective features of nature, are contingent upon the mode of being within which "the natural" as such is disclosed. In other words, technol-

36 Andrew Feenberg, *Questioning Technology* (London: Routledge, 1999), 101.

37 Crucially, Hornborg makes a distinction between a necessary and *sufficient* explanation. Although the physical work it carries out is certainly necessary to explain the function of technology, it is not sufficient. See Alf Hornborg, "Post-Capitalist Ecologies: Energy, 'Value,' and Fetishism in the Anthropocene," *Capitalism Nature Socialism* 24, no. 7 (2016): 71.

38 Alf Hornborg, "Zero-Sum World: Challenges in Conceptualizing Environmental Load Displacement and Ecologically Unequal Exchange in the World-System," *International Journal of Comparative Sociology* 50, nos. 3–4 (2009): 241.

ogy is not merely assisting in nature's self-fulfillment—viewed either as the objective harnessing of the physical regularities of the natural world or as the truthful revelation of nature's inherent potential—but is conditioned by humanity's understanding of itself and its place in the world. Yet, by masquerading as the mere realization of nature, planetary technicity is able to continue its global appropriation of the entire earth as a standing reserve. Because of the naturalization of technology, planetary technicity ensures that we remain mystified by its apparent self-propulsion.

Every Encounter with Wholeness Is Haunted by the Most Terrific Alienation

Another reason for explicitly addressing the earth system paradigm is that our society is the context in which global change research actually takes place, and where its findings are directed toward practical action. The knowledge that earth system scientists produce will go into the public and policy domains, where it will face a number of possible fates: this knowledge can be debated, reconfigured, and developed in the context of other fields of knowledge, and it can be used in decision-making processes. In any case, the concept of the earth system is already deeply embedded in the political institutions that inform our planning for climate adaptation and mitigation and also for many other policy responses to global environmental change. This paradigm shift represents a marked change in our understanding of the correct questions to be asked, the relevant problems to be addressed, and the supposed responsibilities that we face. In order to address these global challenges, earth system scientists have argued the need for new kinds of institutional and operational principles to better equip society for responding to the interconnected challenges we face, which, among other things, includes a synthesis of the knowledge produced across disciplines, "bringing our collective perspective back up to the better rendered 'big picture' of our world. This world is a complex world, and it is inescapable that improving understanding requires the integration of multiple perspectives. If Earth system science is to be a part of this integration process, it requires a much fuller recognition of the interconnectivity of its social and environmental components."[39]

"Interdisciplinary" research is the go-to term in the Anthropocene discourse,[40] but the idea of "the big picture" — which is absolutely crucial to the earth system paradigm — seems rather to suggest that "transdisciplinarity" is an even more accurate description of what is advocated, that is, the same kind of "common formalism" requested by Hans Joachim Schellnhuber.[41] Hence, for earth system scientists, global environmental change is a consequence of a collection of processes — natural and artificial — in *interaction* with each other, such that "the big picture" would provide us with an understanding that is more than merely the sum of its parts. It would thus not be too provocative to venture the claim that global change research is an inherently transdisciplinary idea. Although, as stated by the Committee on the Human Dimensions of Global Change of the National Research Council, "structuring support for human dimensions research only around themes defined by natural science is inadequate,"[42] it is quite obvious upon which premises the social sciences and the humanities are allowed to participate, namely, upon the ontologically flat basis of systems theory.

A central concern, then, is to examine how the assumptions of the current Anthropocene discourse have been set up, what they imply, and how we arrived at them. We have argued herein that one of the main challenges is that the earth system paradigm has set up the terms of its own object of study in such a

39 Cornell et al., "Earth System Science and Society," 6.

40 Michael A. Ellis and Zev M. Trachtenberg, "Which Anthropocene Is It to Be? Beyond Geology to a Moral and Public Discourse," *Earth's Future* 2, no. 2 (2014): 122–25.

41 Frank Oldfield et al., "*The Anthropocene Review:* Its Significance, Implications and the Rationale for a New Transdisciplinary Journal," *Anthropocene Review* 1, no. 1 (2013): 3–7.

42 National Research Council (NRC), *Human Dimensions of Global Environmental Change: Research Pathways for the Next Decade* (Washington, DC: National Academy Press, 1999), 62.

manner that it seeks to understand and predict the properties of a system of which humanity itself is a part, such that humankind's agency is, paradoxically, both extremely powerful in its ability to alter the natural world according to its own will and profoundly powerless in that it is always only a part of the system that it perceives itself to manipulate as if from the outside. This means that, on the one hand, we are led to worry about global environmental change only insofar as we perceive our actions to be those of an external agent, an environmental bandit outside of the geological economy. On the other hand, however, the marks left by us upon the environment, it is argued, are as perfectly natural as the aurora borealis. Given the centrality of humanity in both causing and being affected by global environmental change, we thus face a serious challenge: How do we make sense of the apparent paradox between artifice and nature as it figures in the Anthropocene discourse? It is here that our preceding genealogy instructs us to turn the antinomy into contradiction rather than accepting it at face value: it only appears paradoxical insofar as we assume that the enchantment of technology is incompatible with the instrumental rational attitude. The mythologization that imparts the past, present, and future with numinous qualities is neither a new cultural phenomenon nor one that is exclusively linked to fantasy and affect. Any concrete enactment of an intuitive, sensible, mythical, and, in short, enchanted natural world comes with its own intellectual history that needs to be illuminated if we wish to take the challenge posed by the Anthropocene seriously, and seek to explore how the cybernetic ideology of instrumental mastery has developed over time. Put differently, myths are not only the subject of literature, fine arts, and cinema. On the contrary, mythology can be found everywhere, particularly in the rationalized sphere of modern science. This rich, mythical repertoire consists of narratives that have developed under specific historical and cultural circumstances, and that are now mobilized to make sense of our contemporary globalized mode of experience. The pressing concern for hermeneutics of the kind pursued in this book is that the enchantment of technology cannot be detached from the cultural and historical breadth of its discourse, and thus may mobilize various forms of dogmatism, extremism, exploitation, and colonialism. By adopting a critical attitude, my aim of the genealogical mode of investigation herein has not been to develop upon the vitalist tendencies outlined as a model for understanding technology, but rather to ask what needs to be in place for us to understand technology—and ourselves, dwelling in its midst—as organically networked in the first place. What underlies the symbolic production—through ideas such as complexity, emergence, and self-organization—of this technological enchantment? What are the constitutive elements of the geoscientific disclosure of technology in organicist terms? Such an investigation seeks to avoid writing itself into the Seinsvergessenheit that characterizes the Anthropocene discourse by refusing to take the postulations of the earth system paradigm at face value, instead asking about its conditions of possibility. To understand how planetary technicity bears upon the present moment, we must first trace the conditions for this particular mode of disclosure.

Although the objective of our preceding genealogy has been to examine how questions concerning the essence of technology intersected in modernity with epistemological and methodological concerns within the domain of earth science, by doing so it has in effect also demonstrated how the disclosure of technology in terms of planetary technicity is historically contingent. This is to dispute neither the necessity of industrialization for the emergence of the Anthropocene nor the deleterious effects on nature that this term attempts to capture. But as the genealogy has indicated, industrialization is neither exhaustive nor sufficient as an explanation for the Anthropocene condition. In contrast to theorists of the Anthropocene, such as Hamilton, Grinevald, and Chakrabarty, the intention of this book has been to historicize the Anthropocene not as a natural historical epoch, but rather as the product of a scientific paradigm, and it has attempted to do so by developing a hermeneutic lens that would allow us to turn our investigative eyes to its conditions of possibility. From a genealogical point of view, it is thus no coincidence that Grinevald and Hamilton advance their thesis about the novelty of the Anthropocene by unconsciously consigning precisely what is essential to the current understanding of technology as planetary technicity to certain outdated, historical beliefs. Because the notion of a break with the past and the repetitive assertion that we now live in a "post" era — whether it be "postmodern," "postideological," "postpolitical," "posthuman," "postnatural," and so on — is precisely what such a genealogical method aims to contest.[43] Hence, by proceeding genealogically, I have sought to historicize an imagination that has become all the more important as it has disseminated and faded away, that is, as it has unreflexively seeped into the background so as to become the present-at-hand condition of our most basic experience of the earth as an ontologically flat system — an earth system. For it is one thing to historically examine the spread and use of metaphors and images of the earth, and it is one thing to account for the uneasy convergence of remote sensing with a global vocabulary by resituating the history of technological change in a broad context concerning the relationship between nature and artifice in the modern age, but it is something else entirely to track the horizon that discloses the earth as a system in the first place, and that, anterior to the writing of any history of the synthetic hybridization of nature and artifice, has come to dictate our historical experience of this ontological collapse.[44] Thus the stage has now been set for the possibility to critically reflect on what it means to inhabit a world in which planetary technicity has historically emerged and has become essential, or, put differently, in which it has set up the global enframing of the earth as a self-organizing system, thus naturalizing technology

---

43 In the words of the sociologist Luigi Pellizzoni, "The issue, [...] at least from a genealogical perspective, is not so much to assess how innovative [certain] ontologies are [...] as to understand the role such ontologies play in the present historical juncture"; Pellizzoni, "New Materialism and Runaway Capitalism: A Critical Assessment," *Soft Power* 5, no. 1 (2017): 68–69.

44 Benjamin Lazier, "Earthrise; or, the Globalization of the World Picture," *American Historical Review* 116, no. 3 (2011): 626–27.

in such a manner that it comes to act upon us in ways we do not even notice as "technological" any longer. To be sure, on our genealogical journey, we have been faced with numerous examples of environmental awareness, ecological concern for nature, and even feelings of mystical communion with the earth. Yet, we have also indicated that for every impulse to care, there are injunctions to manage and control, insofar as every encounter with wholeness is haunted by the most terrific alienation.[45]

Once Technology Takes Root

If there is anything like a main message of this entire book, then it is this: that nature, in the Anthropocene discourse, has become entirely functional in relation to the technical systems that now enframe and construct our place of dwelling in terms of hybrid environments — what Jean Baudrillard, the French sociologist and diagnostician of the postmodern condition, described as an infusion of the virtual into the actual and an absorption and appropriation of the actual into the virtual, a process that is at once ubiquitous and naturalized for us in our simulacrum of artificial worlds.[46] The peculiar problem is that we no longer see or notice this kind of technological appropriation of nature, because as a result of the collapse of the boundary between the natural and the artificial, we lack the critical distance that critique requires in order to acknowledge our being trapped in the meshes of the latter. At the point that nature entirely disappears, there thus transpires a simultaneous *Bodenständigkeit* — rendering technology "down to earth" or "rooted in the soil" — in the disclosure of what is nevertheless a highly technologized globe, which results in a rather peculiar undecidability between nature and artifice. With a little help from Jameson, we might say this:

45 Ibid., 630.

46 Jean Baudrillard, *Symbolic Exchange and Death,* trans. Iain Hamilton Grant (London: Sage, 1993), 31.

> To do away with the last remnants of nature and with the natural as such is surely the secret dream and longing of [the Anthropocene] — even though it is a dream the latter dreams with the secret proviso that "nature" never really existed in the first place anyhow. This is then the moment at which it becomes obligatory to observe that [the Anthropocene] is also the moment of a host of remarkable and dramatic "revivals" of nature — [... and w]hat can lie beyond what Marx called *naturwüchsige* modes of production, if not simply more capitalism albeit of a more technologically sophisticated and globalized variety?[47]

Ascribing ontological primacy to the technological as a means of challenging the dominance of instrumental reason is thus to end up complicit in fabricating, in place of the modern cogito, a sense of selfhood that is, as a matter of fact, perfectly functional in relation to the logic of late capitalism. In the intellectual historian Susan Buck-Morss's words:

> One could say that the dynamics of capitalist industrialism [has] caused a curious reversal in which "reality" and "art" switch places. Reality becomes artificial, a phantasmagoria of commodities and architectural construction made possible by the new industrial processes. The modern city [is] nothing but the proliferation of such objects, the density of which created an artificial landscape of buildings and consumer items as totally encompassing as the earlier, natural one.[48]

In the twenty-first century, then, we must rather make sense of this puzzling inversion whereby technology has been attributed the same organic features that many of its twentieth-century critics prescribed as an antidote to the modern condition: tech-

47 Fredric Jameson, *The Seeds of Time* (New York: Columbia University Press, 1996), 46, 47.

48 Susan Buck-Morss, "Benjamin's *Passagen-Werk*: Redeeming Mass Culture for the Revolution," *New German Critique* 29 (1983): 213.

nology, in the Anthropocene, is disclosed precisely as the poietic expression of nature's self-organization. It is the apprehension connected to this simultaneous covering and stripping bare of the globe that allows itself to be expressed so effectively in the Anthropocene discourse. With the globalization of technology, it is no longer possible to identify where the artificial starts and the natural ends, since the same process of anthropogenic manipulation that slowly erodes the resilience of nature also erodes the very boundaries of the artificial as a category.

It is in this manner that technology may spread and extend across the globe all the way to the point where, as it culminates in the power of a geophysical force, it loses its properties as artificial and, with the loss of these properties, that which would allow it to be distinguished from nature in the first place. Hence, planetary technicity does not "intervene" in nature as if from the outside, rather, it names the complete artificialization of nature itself, to the point of an indistinguishability between nature and artifice — what the philosopher Jean-Luc Nancy has called "the *ecotechnology* that our ecologies and economies have already become."[49] Nancy writes: "Our world is the world of the 'technical,' a world whose cosmos, nature, gods, entire system is, in its inner joints, exposed as 'technical'[. … F]or the projections of linear histories and final ends," planetary technicity "substitutes the spacings of time, local differences, and numerous bifurcations. [It] deconstructs the system of ends, renders them unsystemizable, nonorganic, even stochastic (*except* through an imposition of the ends of political economy or capital)."[50] What remains is an endless proliferation that changes means to ends indefinitely without ever progressing anywhere.[51] The only available cohesion is therefore that of an endless reordering of beings

49 Jean-Luc Nancy and Aurelien Barrau, *What's These Worlds Coming To?*, trans. Travis Holloway and Flor Méchain (New York: Fordham University Press, 2015), 54.

50 Jean-Luc Nancy, *Corpus*, trans. Richard A. Rand (New York: Fordham University Press, 2008), 89.

51 Susanna Lindberg, "Onto-Technics in Bryant, Harman, and Nancy," *PhaenEx* 12, no. 2 (Winter 2018): 95–96.

within a framework where no single order is valued higher than any other order. Contingency, fortuity, and juxtaposition are the tenets of this logic: a permanent reconfiguration that neither responds nor tends toward any higher meaning. Once the enframing of technology holds sway globally, every parcel of sense relaunches the experience of the world onto limitlessness: planetary technicity eliminates anything that would stand in the way of the endless expansion that is so necessary for the capitalist mode of production. This is a world of endless growth that makes no sense other than its own self-augmentation.

Accordingly, it is only in the critical vein of Kant's *non-naturalization* of our discursive frame that we may grasp the naturalization of technology as a symptom of the technification of nature, or, as the philosopher Luciano Floridi puts it, to account for "why science is increasingly artefactual" while "the naturalisation of the non-natural turns out to be an expression of the artefactual nature of the natural."[52] It is solely through elaboration of and negotiation with the unescapably *artifactual* aspects of our discursive frame that we come to better grasp the natural world independently of said framework, and from which we may thus advance a project of proper naturalization. Only by denaturalizing planetary technicity do we consequently naturalize it, in the slightly counterintuitive sense that a proper naturalization consists of a refusal to accept either "the natural" or "the artificial" as naturalistic. Floridi notes:

> This leads to the conclusion that our interpretation of the naturalization of phenomena, including cultural and philosophical ones, is itself a cultural phenomenon, and hence non-naturalistic. It is the non-natural that enables us to create categories such as "natural," "naturalized," and "naturalization." There is nothing anti-realist or relativistic in such an acknowledgement. It only means that such a construction would not be reducible to a natural process without informa-

52 Luciano Floridi, "A Plea for Non-Naturalism as Constructionism," *Minds & Machines* 27 (2017): 283, 284.

> tion loss[. ... T]he non-natural is our first nature, and the natural is actually our second nature. And this means that what we need is a genealogy of the natural from the non-natural, not vice versa.[53]

Already the Romantics were correct to point out the absurdity and danger of the dominance of substance dualism in the modern epoch, but it is now necessary to revisit and reevaluate the monistic hybridism that has slowly begun to take its place. For "if our understanding of technology pre-emptively dissolves the difference between original and copy, life and syntax, nature and its mimesis,"[54] then technology itself will appear as but an expression of a self-organizing nature. The danger is that technology can thus all too easily appear as a natural product, as if directed from within—just like the enchanted fetish—and such an understanding would risk rendering it identical to the earth itself.

53 Ibid., 283.

54 Anson Rabinbach, quoted in Andrew Bowie, "Romanticism and Technology," *Radical Philosophy* 72 (1995): 10.

Bibliography

Adams, Frank D. *The Birth and Development of the Geological Sciences*. Baltimore: The Williams and Wilkins Company, 1938.

Althusser, Louis. *Philosophy and the Spontaneous Philosophy of the Scientists & Other Essays*. Edited by Gregory Elliott. Translated by Ben Brewster, James H. Kavanagh, Thomas E. Lewis, Grahame Lock, and Warren Montag. London: Verso, 1990.

Arendt, Hannah. *The Human Condition*. Chicago: University of Chicago Press, 1958. DOI: 9780226586748.001.0001.

———. "Man's Conquest of Space." *American Scholar* 32, no. 4 (1963): 527–40. https://www.jstor.org/stable/41209127.

Aristotle. *Aristotle's "Physics."* Edited by William D. Ross. Oxford: Clarendon, 1936. DOI: 10.1093/actrade/9780198141099.book.1.

———. *The Metaphysics, Books I–IX*. Translated by Hugh Tredennick. London: William Heinemann Ltd., 1933.

Ashby, William R. "Principles of the Self-Organizing System." In *Principles of Self-Organization: Transactions of the University of Illinois Symposium of Self-Organization, 8 and 9 June, 1961*, edited by Heinz von Förster and George W. Zopf, 255–78. New York: Pergamon Press, 1962.

Bailes, Kendall, E. *Science and Russian Culture in an Age of Revolutions: V.I. Vernadsky and His Scientific School, 1863–1945*. Bloomington: Indiana University Press, 1990.

Bailey, Edward B. *James Hutton: The Founder of Modern Geology*. Amsterdam: Elsevier, 1967.

Barnosky, Anthony D., Elizabeth A. Hadly, Jordi Bascompte, Eric L. Berlow, James H. Brown, Mikael Fortelius, Wayne M. Getz, John Harte, Alan Hastings, Pablo A. Marquet, Neo D. Martinez, Arne Mooers, Peter Roopnarine, Geerat Vermeij, John W. Williams, Rosemary Gillespie, Justin Kitzes, Charles Marshall, Nicholas Matzke, David P. Mindell, Eloy Revilla, and Adam B. Smith. "Approaching a State Shift in Earth's Biosphere." *Nature* 486, no. 7401 (2012): 52–58. DOI: 10.1038/nature11018.

Anthony D., Barnosky, Nicholas Matzke, Susumu Tomiya, Guinevere O.U. Wogan, Brian Swartz, Tiago B. Quental, Charles Marshall, Jenny L. McGuire, Emily L. Lindsey, Kaitlin C. Maguire, Ben Mersey, and Elizabeth A. Ferrer. "Has the Earth's Sixth Mass Extinction Already Arrived?" *Nature* 471, no. 7336 (2011): 51–57. DOI: 10.1038/nature09678.

Barthes, Roland. *Mythologies*. Translated by Annette Lavers. New York: Noonday Press, 1991.

Baskin, Jeremy. "Paradigm Dressed as Epoch: The Ideology of the Anthropocene." *Environmental Values* 24 (2015): 9–29. DOI: 10.3197/096327115X14183182353746.

Bates, David R., and Marcel Nicolet. "Atmospheric Hydrogen." *Planetary and Space Science* 13, no. 9 (1965): 905–9. DOI: 10.1016/0032-0633(65)90175-3.

Bateson, Gregory. *Mind and Nature: A Necessary Unity*. New York: E.P. Dutton, 1979.

———. *Steps to an Ecology of Mind: Collected Essays in Anthropology, Psychiatry, Evolution, and Epistemology*. London: Northvale, 1987.

———. "This Normative Natural History Called Epistemology." In *A Sacred Unity: Further Steps to an Ecology of Mind,* edited by Rodney E. Donaldson, 215–24. New York: Harper, 1991.

Bateson, Gregory, and Mary C. Bateson. *Angels Fear: Towards an Epistemology of the Sacred.* New York: Bantam Books, 1988.

Baudrillard, Jean. *Symbolic Exchange and Death.* Translated by Iain Hamilton Grant. London: Sage, 1993. DOI: 10.4135/9781446280423.

Bennett, Jane. *Vibrant Matter: A Political Ecology of Things.* Durham: Duke University Press, 2010. DOI: 10.1215/9780822391623.

Bloch, Ernst. *The Principle of Hope.* 3 Vols. Oxford: Blackwell, 1986.

Blok, Vincent. "Earthing Technology: Towards an Eco-Centric Concept of Biomimetic Technologies in the Anthropocene." *Techné: Research in Philosophy and Technology* 21, nos. 2–3 (2017): 127–49. DOI: 10.5840/techne201752363.

Bonneuil, Christophe, and Jean-Baptiste Fressoz. *The Shock of the Anthropocene: The Earth, History and Us.* London: Verso, 2016.

Borges, Jorge L. "The Fearful Sphere of Pascal." In *Labyrinths: Selected Stories & Other Writings*, 189–92. New York: New Directions, 1962.

Bowie, Andrew. "Romanticism and Technology." *Radical Philosophy* 72 (1995): 5–16. https://www.radicalphilosophy.com/article/romanticism-and-technology.

Braje, Todd J., and Jon M. Erlandson. "Looking Forward, Looking Back: Humans, Anthropogenic Change, and the Anthropocene." *Anthropocene* 4 (2013): 116–21. DOI: 10.1016/j.ancene.2014.05.002.

Brandom, Robert B. *Articulating Reasons: An Introduction to Inferentialism.* Cambridge: Harvard University Press, 2001. DOI: 10.4159/9780674028739.

Bratton, Benjamin H. *The Terraforming.* Moscow: Strelka Press, 2019.

Braudel, Fernand. *The Structures of Everyday Life*, Vol. 1: *Civilization & Capitalism, 15th–18th Century.* New York: HarperCollins, 1982.

Brito, Lidia, and Mark Stafford Smith. *State of the Planet Declaration — Planet under Pressure: New Knowledge towards Solutions Conference, London, 26–29th of March 2012*. London: Diversitas, 2012.

Bryant, William H. "Whole System, Whole Earth: The Convergence of Technology and Ecology in Twentieth-Century American Culture." PhD diss., University of Iowa, 2006.

Buck-Morss, Susan. "Benjamin's *Passagen-Werk*: Redeeming Mass Culture for the Revolution." *New German Critique* 29 (1983): 211–40. DOI: 10.2307/487795.

Bukatman, Scott. *Terminal Identity: The Virtual Subject in Postmodern Science Fiction*. Durham: Duke University Press, 1993. DOI: 10.1515/9780822379287.

Carlyle, Thomas. "Boswell's Life of Johnson." In *Macaulay's and Carlyle's Essays on Samuel Johnson*, edited by William Strunk, Jr., 65–158. New York: Henry Holt, 1895.

Cera, Agostino. "The Technocene or Technology as (Neo) Environment." *Techné: Research in Philosophy and Technology* 21, nos. 2–3 (2017): 243–81. DOI: 10.5840/techne201710472.

Chakrabarty, Dipesh. "Anthropocene Time." *History & Theory* 57, no. 1 (2018): 5–32. DOI: 10.1111/hith.12044.

———. "Postcolonial Studies and the Challenge of Climate Change." *New Literary History* 43, no. 1 (2012): 1–18. DOI: 10.1353/nlh.2012.0007.

———. "The Climate of History: Four Theses." *Critical Inquiry* 35, no. 2 (2009): 197–222. DOI: 10.1086/596640.

Cixous, Hélène. "Fiction and Its Phantoms: A Reading of Freud's Das Unheimliche (the 'Uncanny')." *New Literary History* 7, no. 3 (1976): 525–48, 619–45. DOI: 10.2307/468561.

Clark, Nigel. "Earth, Fire, Art: Pyrotechnology and the Crafting of the Social." In *Inventing the Social*, edited by Noortje Marres, Michael Guggenheim, and Alex Wilkie, 173–94. Manchester: Mattering Press, 2018. DOI: 10.1353/book.81374.

———. "Fiery Arts: Pyrotechnology and the Political Aesthetics of the Anthropocene." *GeoHumanities* 1, no. 2 (2015): 266–84. DOI: 10.1080/2373566X.2015.1100968.
———. *Inhuman Nature: Sociable Life on a Dynamic Planet.* London: SAGE, 2011. DOI: 10.4135/9781446250334.
———. "Rock, Life, Fire: Speculative Geophysics and the Anthropocene." *Oxford Literary Review* 34, no. 2 (2012): 259–76. DOI: 10.3366/olr.2012.0045.
Clarke, Bruce. "Autopoiesis and the Planet." In *Impasses of the Post-Global: Theory in the Era of Climate Change,* Vol. 2, edited by Henry Sussman, 60–77. Ann Arbor: Open Humanities Press, 2012. DOI: 10.3998/ohp.10803281.0001.001.
———. *Gaian Systems: Lynn Margulis, Neocybernetics, and the End of the Anthropocene.* Minneapolis: University of Minnesota Press, 2020. DOI: 10.5749/j.ctv16f6d9c.
———. "Neocybernetics of Gaia: The Emergence of Second-Order Gaia Theory." In *Gaia in Turmoil: Climate Change, Biodepletion, and Earth Ethics in an Age of Crisis,* edited by Eileen Crist and H. Bruce Rinker, 293–314. Cambridge: MIT Press, 2009. DOI: 10.7551/mitpress/7845.001.0001.
Clynes, Manfred E., and Nathan S. Kline. "Cyborgs and Space." In *The Cyborg Handbook,* edited by Chris H. Gray, Heidi J. Figueroa-Sarriera, and Steven Mentor, 29–34. London: Routledge, 1995.
Connes, Janine, and Pierre Connes. "Near-Infrared Planetary Spectra by Fourier Spectroscopy. I. Instruments and Results." *Journal of the Optical Society of America* 56, no. 7 (1966): 896–910. DOI: 10.1364/JOSA.56.000896.
Conty, Arianne. "Who Is to Interpret the Anthropocene? Nature and Culture in the Academy." *La Deleuziana* 4 (2016): 19–44. http://www.ladeleuziana.org/wp-content/uploads/2017/06/Conty-%E2%80%93-Who-is-to-Interpret-the-Anthropocene.pdf.
Cooper, Melinda. *Life as Surplus: Biotechnology and Capitalism in the Neoliberal Era.* Seattle: University of Washington Press, 2008.

———. "Life, Autopoiesis, Debt: Inventing the Bioeconomy." *Distinktion: Journal of Social Theory* 8, no. 1 (2007): 25–43. DOI: 10.1080/1600910X.2007.9672937.

Cornell, Sarah E., Catherine J. Downy, Evan D.G. Fraser, and Emily Boyd. "Earth System Science and Society: A Focus on the Anthroposphere." In *Understanding the Earth System: Global Change Science for Application,* edited by Sarah E. Cornell, I. Colin Prentice, Joanna I. House, and Catherine J. Downy, 1–38. Cambridge: Cambridge University Press, 2012. DOI: 10.1017/CBO9780511921155.004.

Coxon, Allan H. *The Fragments of Parmenides: A Critical Text with Introduction and Translation, the Ancient Testimonia and a Commentary.* Translated by Richard McKirahan. Las Vegas: Parmenides Publishing, 2009.

Crutzen, Paul J. "Geology of Mankind." *Nature* 415, no. 6867 (2002): 23. DOI: 10.1038/415023a.

Crutzen, Paul J., and Eugene F. Stoermer. "The 'Anthropocene.'" *IGBP Global Change Newsletter* 41 (2000): 17–18. http://www.igbp.net/download/18.316f18321323470177580001401/1376383088452/NL41.pdf.

Darwin, Charles. *On the Origin of Species by Means of Natural Selection, or the Preservation of Favored Races in the Struggle for Life.* London: John Murray, 1859.

Dawkins, Richard. *The Blind Watchmaker: Why the Evidence of Evolution Reveals a Universe without Design.* New York: W.W. Norton, 1986.

———. *The Extended Phenotype: The Gene as the Unit of Selection.* Oxford: Oxford University Press, 1982.

———. *The Selfish Gene.* Oxford: Oxford University Press, 1976.

———. *Unweaving the Rainbow: Science, Delusion, and the Appetite for Wonder.* Boston: Houghton Mifflin, 2000.

Dean, Dennis R. *James Hutton and the History of Geology.* Ithaca: Cornell University Press, 1992. DOI: 10.7591/9781501733994.

Delanda, Manuel. *Assemblage Theory.* Edinburgh: Edinburgh University Press, 2016. DOI: 10.1515/9781474413640.

Delio, Ilia. "Transhumanism or Ultrahumanism? Teilhard de Chardin on Technology, Religion, and Evolution." *Theology & Science* 10, no. 2 (2012): 153–66. DOI: 10.1080/14746700.2012.669948.
Dennis, Michael A. "Earthly Matters: On the Cold War and the Earth Sciences." *Social Studies of Science* 33, no. 5 (2003): 809–19. DOI: 10.1177/0306312703335007.
Dirzo, Rodolfo, Hillary S. Young, Mauro Galetti, Gerardo Ceballos, Nick J.B. Isaac, and Ben Collen. "Defaunation in the Anthropocene." *Science* 345, no. 6195 (2014): 401–6. DOI: 10.1126/science.1251817.
Dobzhansky, Theodosius. "Nothing in Biology Makes Sense Except in the Light of Evolution." *American Biology Teacher* 35, no. 3 (1973): 125–29. DOI: 10.2307/4444260.
Doel, Ronald E. "Constituting the Postwar Earth Sciences: The Military's Influence on the Environmental Sciences in the USA after 1945." *Social Studies of Science* 33, no. 5 (2003): 635–66. DOI: 10.1177/0306312703335002.
Donovan, Arthur, and Joseph Prentiss. "James Hutton's Medical Dissertation." *Transactions of the American Philosophical Society* 70, no. 6 (1980): 3–57. DOI: 10.2307/1006364.
Doolittle, W. Ford. "Is Nature Really Motherly?" *Co-Evolution Quarterly* (Spring 1981): 58–63.
Dukes, Hunter. "Assembling the Mechanosphere: Monod, Althusser, Deleuze, and Guattari." *Deleuze Studies* 10, no. 4 (2016): 514–30. DOI: 10.3366/dls.2016.0243.
Dupuy, Jean-Pierre. "Cybernetics Is an Antihumanism: Technoscience and the Rebellion Against the Human Condition." In *French Philosophy of Technology Classical Readings and Contemporary Approaches,* edited by Sascha Loeve, Xavier Guchet, and Bernadette Bensaude-Vincent, 139–56. Berlin: Springer, 2018. DOI: 10.1007/978-3-319-89518-5_9.
Edwards, Paul N. *The Closed World Computers and the Politics of Discourse in Cold War America.* Cambridge: MIT Press, 1997.

Ehrlich, Paul R., John Harte, Mark A. Harwell, Peter H. Raven, Carl Sagan, George M. Woodwell, Joseph Berry, Edward S. Ayensu, Anne H. Ehrlich, Thomas Eisner, Stephen J. Gould, Herbert D. Grover, Rafael Herrera, Robert M. May, Ernst Mayr, Christopher P. McKay, Harold A. Mooney, Norman Myers, David Pimentel, and John M. Teal. "Long-Term Biological Consequences of Nuclear War." *Science* 222, no. 1 (1983): 1293–300. DOI: 10.1126/science.6658451.

Elden, Stuart. "Missing the Point: Globalization, Deterritorialization, and the Space of the World." *Transactions of the Institute of British Geographers* 30, no. 1 (2005): 8–19. DOI: 10.1111/j.1475-5661.2005.00148.x.

Ellis, Erle C. "Ecology in an Anthropogenic Biosphere." *Ecological Monographs* 85, no. 3 (2015): 287–331. DOI: 10.1890/14-2274.1.

———. "The Planet of No Return." In *Love Your Monsters: Postenvironmentalism and the Anthropocene,* edited by Michael Schellenberger and Ted Nordhaus, 37–46. Oakland: The Breakthrough Institute, 2011.

Ellis, Erle C., and Peter K. Haff. "Earth Science in the Anthropocene: New Epoch, New Paradigm, New Responsibilities." *EOS Transactions* 90 (2009): 473. DOI: 10.1029/2009EO490006.

Ellis, Erle C., and Navin Ramankutty. "Putting People in the Map: Anthropogenic Biomes of the World." *Frontiers in Ecology and the Environment* 6, no. 8 (2008): 439–47. DOI: 10.1890/070062.

Ellis, Michael A., and Zev M. Trachtenberg. "Which Anthropocene Is It to Be? Beyond Geology to a Moral and Public Discourse." *Earth's Future* 2, no. 2 (2014): 122–25. DOI: 10.1002/2013EF000191.

Ellul, Jacques. *The Technological System.* Translated by Joachim Neugroschel. New York: Continuum, 1980.

Engels, Friedrich. "Letter to Pyotr Lavrov, 12–17 November 1875." Translated by Peter Ross and Betty Ross. In *The Collected Works of Karl Marx and Friedrich Engels,* Vol. 45: *Letters 1874–1879,* edited by Jack Cohen, Maurice Cornforth,

Maurice Dobb, Eric J. Hobsbawm, James Klugmann, and Margaret Mynatt, 106–9. London: Lawrence & Wishart, 2010.

Falkowski, Paul, Robert J. Scholes, Edward A. Boyle, Josep Canadell, Donald Canfield, James Elser, Nicolas Gruber, Ken Hibbard, Peter Högberg, S. Linder, F.T. Mackenzie, B. Moore III, T. Pedersen, Y. Rosenthal, S. Seitzinger, V. Smetacek, and W. Steffen. "The Global Carbon Cycle: A Test of Our Knowledge of Earth as a System." *Science* 290, no. 5490 (2000): 291–96. DOI: 10.1126/science.290.5490.291.

Feenberg, Andrew. *Questioning Technology.* London: Routledge, 1999. DOI: 10.4324/9780203022313.

Ffytche, Matt. *The Foundation of the Unconscious: Schelling, Freud, and the Birth of the Modern Psyche.* Cambridge: Cambridge University Press, 2012. DOI: 10.1017/CBO9781139024006.

Fleming, James R. *Fixing the Sky: The Checkered History of Weather and Climate Control.* New York: Columbia University Press, 2010.

Floridi, Luciano. "A Plea for Non-Naturalism as Constructionism." *Minds & Machines* 27 (2017): 269–85. DOI: 10.1007/s11023-017-9422-9.

Förster, Heinz von. "On Self-Organizing Systems and Their Environments." In *Understanding Understanding: Essays on Cybernetics and Cognition,* 1–19. Berlin: Springer, 2002. DOI: 10.1007/0-387-21722-3_1.

Fraser, Evan D.G., Andrew J. Dougill, Klaus Hubacek, Claire H. Quinn, Jan Sendzimir, and Mette Termansen. "Assessing Vulnerability to Climate Change in Dryland Livelihood Systems: Conceptual Challenges and Interdisciplinary Solutions." *Ecology and Society* 16, no. 3 (2011): 1–12. DOI: 10.5751/ES-03402-160303.

Frye, Northrop. "The Drunken Boat: The Revolutionary Element in Romanticism." In *Romanticism Reconsidered,* edited by Northrop Frye, 1–25. New York: Columbia University Press, 1963.

Furniss, Tom. "James Hutton's Geological Tours of Scotland: Romanticism, Literary Strategies, and the Scientific Quest." *Science & Education* 23, no. 3 (2014): 565–88. DOI: 10.1007/s11191-012-9464-6.

Gacheva, Anastasia. "Art as the Overcoming of Death: From Nikolai Fedorov to the Cosmists of the 1920s." *e-flux* 89 (2018): 1–12. https://www.e-flux.com/journal/89/180332/art-as-the-overcoming-of-death-from-nikolai-fedorov-to-the-cosmists-of-the-1920s/.

Galison, Peter. "The Ontology of the Enemy: Norbert Wiener and the Cybernetic Vision." *Critical Inquiry* 21, no. 1 (1994): 228–66. DOI: 10.1086/448747.

Galloway, James N., Frank J. Dentener, D.G. Capone, Elizabeth W. Boyer, Robert W. Howarth, Sybil P. Seitzinger, Gregory P. Asner, Cory C. Cleveland, Pamela A. Green, Elizabeth A. Holland, D.M. Karl, A.F. Michaels, J.H. Porter, A.R. Townsend, and C.J. Vöosmarty. "Nitrogen Cycles: Past, Present, and Future." *Biogeochemistry* 70, no. 2 (2004): 153–226. DOI: 10.1007/s10533-004-0370-0.

German Advisory Council on Global Change (WBGU). *World in Transition: A Social Contract for Sustainability. Flagship Report 2011.* Berlin: WBGU, 2011.

Gilson, Étienne. *From Aristotle to Darwin and Back Again: A Journey in Final Causality, Species, and Evolution.* Translated by John Lyon. Notre Dame: University of Notre Dame Press, 1984. DOI: 10.2307/j.ctvpj79v6.

Ginn, Franklin, Michelle Bastian, David Farrier, and Jeremy Kidwell. "Introduction: Unexpected Encounters with Deep Time." *Environmental Humanities* 10, no. 1 (2018): 213–25. DOI: 10.1215/22011919-4385534.

Gould, Stephen J. *Time's Arrow, Time's Cycle: Myth and Metaphor in the Discovery of Geological Time.* Cambridge: Harvard University Press, 1987.

Grant, Iain Hamilton "Mining Conditions: A Response to Harman." In *The Speculative Turn: Continental Materialism and Realism,* edited by Levi Bryant, Nick Srnicek, and Graham Harman, 41–46. Melbourne: re.press, 2011.

Gray, Donald, P. "Teilhard's Energy." *CrossCurrents* 21, no. 2 (1971): 238–40.

Grinevald, Jacques. "On a Holistic Concept for Deep and Global Ecology: The Biosphere." *Fundamenta Scientiae* 8, no. 2 (1987): 197–226.

Grübler, Arnulf. *Technology and Global Change.* Cambridge: University of Cambridge Press, 1998. DOI: 10.1017/CBO9781316036471.

Guillaume, Bertrand. "Vernadsky's Philosophical Legacy: A Perspective from the Anthropocene." *Anthropocene Review* 1, no. 2 (2014): 137–46. DOI: 10.1177/2053019614530874.

Haff, Peter K. "Being Human in the Anthropocene." *Anthropocene Review* 4, no. 2 (2017): 103–9. DOI: 10.1177/2053019617700875.

———. "Humans and Technology in the Anthropocene: Six Rules." *Anthropocene Review* 1, no. 2 (2014): 126–36. DOI: 10.1177/2053019614530575.

———. "Technology as a Geological Phenomenon: Implications for Human Well-Being." *Geological Society Special Publication* 395, no. 1 (2013): 301–9. DOI: 10.1144/SP395.4

Hamilton, Clive. *Defiant Earth: The Fate of Humans in the Anthropocene.* Cambridge: Polity Press, 2017.

———. "Define the Anthropocene in Terms of the Whole Earth." *Nature* 536, no. 7616 (2016): 251. DOI: 10.1038/536251a.

———. "The Theodicy of the 'Good Anthropocene.'" *Environmental Humanities* 7, no. 1 (2015): 233–38. DOI: 10.1215/22011919-3616434.

Hamilton, Clive, Christophe Bonneuil, and François Gemenne. "Thinking the Anthropocene." In *The Anthropocene and the Global Environmental Crisis: Rethinking Modernity in a New Epoch,* edited by Clive Hamilton, Christophe Bonneuil, and François Gemenne, 1–14. London: Routledge, 2015. DOI: 10.4324/9781315743424-1.

Hamilton, Clive, and Jacques Grinevald. "Was the Anthropocene Anticipated?" *Anthropocene Review* 2, no. 1 (2015): 59–72. DOI: 10.1177/2053019614567155.

Haraway, Donna J. "The Biopolitics of Postmodern Bodies: Constitutions of Self in Immune System Discourse." In *Simians, Cyborgs, and Women: The Reinvention of Nature,* 203–30. London: Routledge, 1991. DOI: 10.4324/9780203873106.

———. "Cyborgs and Symbionts: Living Together in the New World Order." In *The Cyborg Handbook,* edited by Chris H. Gray, Heidi J. Figueroa-Sarriera, and Steven Mentor, xi–xx. London: Routledge, 1995.

Hardin, Garrett. "The Tragedy of the Commons." *Science* 162, no. 3859 (1968): 1243–48. https://www.jstor.org/stable/1724745.

Harper, Kristine C. *Weather by the Numbers: The Genesis of Modern Meteorology.* Cambridge: MIT Press, 2008. DOI: 10.7551/mitpress/9780262083782.001.0001.

Havel, Václav. "The Need for Transcendence in the Postmodern World." *The Futurist* (July–August 1995): 46–49.

Hegel, Georg W.F. *Hegel's Philosophy of Nature.* Vol. 1. Edited and translated by Michael J. Petry. London: Allen and Unwin, 1970.

Heidegger, Martin. *Being and Time.* Edited by Dennis J. Schmidt. Translated by Joan Stambaugh. New York: SUNY Press, 2010.

———. "The End of Philosophy and the Task of Thinking." In *Basic Writings,* edited by David F. Krell, translated by Joan Stambaugh, 431–49. New York: HarperCollins, 1978.

———. "Letter on 'Humanism.'" In *Basic Writings,* edited by David F. Krell, 217–65. New York: HarperCollins, 1978.

———. "'Only a God Can Save Us': The *Spiegel* Interview." In *Heidegger: The Man and the Thinker,* edited by Thomas Sheehan, translated by William J. Richardson, 45–68. London: Transaction Publishers, 1981.

———. "The Age of the World Picture." In *The Question Concerning Technology and Other Essays,* translated by William Lovitt, 115–54. New York: Harper & Row, 1977.

———. "The Question concerning Technology." In *The Question concerning Technology and Other Essays,* translated by William Lovitt, 3–35. New York: Harper & Row, 1977.

Heringman, Noah. *Romantic Rocks, Aesthetic Geology.* Ithaca: Cornell University Press, 2004.

Heymann, Matthias, Henrik Knudsen, Maiken L. Lolck, Henry Nielsen, Kristian H. Nielsen, and Christopher J. Ries. "Exploring Greenland: Science and Technology in Cold War Settings." *Scientia Canadensis* 33, no. 2 (2010): 11–42. DOI: 10.7202/1006149ar.

Hitchcock, Dian R., Peter Fellgett, Janine Connes, Pierre Connes, Lewis D. Kaplan, J. Ring, and James E. Lovelock. "Detecting Planetary Life from Earth." *Science Journal* 3 (1967): 56–67.

Höhler, Sabine. *Spaceship Earth in the Environmental Age, 1960–1990.* London: Routledge, 2015. DOI: 10.4324/9781315653921.

Hoły-Łuczaj, Magdalena, and Vincent Blok. "How to Deal with Hybrids in the Anthropocene? Towards a Philosophy of Technology and Environmental Philosophy 2.0." *Environmental Values* 28, no. 3 (2019): 325–46. DOI: 10.3197/096327119X15519764179818.

Horkheimer, Max, and Theodor W. Adorno. *Dialectic of Enlightenment: Philosophical Fragments.* Edited by Gunzelin Schmid Noerr. Translated by Edmund Jephcott. Stanford: Stanford University Press, 2002. DOI: 10.1515/9780804788090.

Hörl, Erich. "Erich Hörl: A *continent.* Inter-view." *continent.* 5, no. 2 (2016): 25–30. https://continentcontinent.cc/archives/issues/issue-5-2-2016/erich-hoerl.

Hornborg, Alf. "Artifacts Have Social Consequences, Not Agency: Toward A Critical Theory of Global Environmental History." *European Journal of Social Theory* 20, no. 1 (2017): 95–110. DOI: 10.1177/1368431016640536.

———. "Post-Capitalist Ecologies: Energy, 'Value,' and Fetishism in the Anthropocene." *Capitalism*

Nature Socialism 24, no. 7 (2016): 61–76. DOI: 10.1080/10455752.2016.1196229.

———. *The Power of the Machine: Global Inequities of Economy, Technology, and Environment.* New York: Altamira Press, 2001.

———. "Zero-Sum World: Challenges in Conceptualizing Environmental Load Displacement and Ecologically Unequal Exchange in the World-System." *International Journal of Comparative Sociology* 50, nos. 3–4 (2009): 237–62. DOI: 10.1177/0020715209105141.

Hui, Yuk. "Machine and Ecology." *Angelaki: Journal of the Theoretical Humanities* 25, no. 4 (2020): 54–66. DOI: 10.1080/0969725X.2020.1790835.

Hume, David. *Enquiries Concerning Human Understanding and Concerning the Principles of Morals.* Oxford: Clarendon, 1902.

Hutchinson, G. Evelyn. "The Biochemistry of the Terrestrial Atmosphere." In *The Earth as a Planet,* edited by Gerard P. Kuiper, 371–433. Chicago: University of Chicago Press, 1954.

———. "The Biosphere." *Scientific American* 223, no. 3 (1970): 44–53. DOI: 10.1038/scientificamerican0970-44.

Hutton, James. *Abstract of a Dissertation Read in the Royal Society of Edinburgh, Upon the Seventh of March, and Fourth of April, MDCCLXXXV, Concerning the System of the Earth, Its Duration, and Stability.* Edinburgh: Royal Society of Edinburgh, 1785.

———. *A Dissertation upon the Philosophy of Light, Heat, and Fire.* Edinburgh: Cadell, Junior & Davies, 1794.

———. *An Investigation of the Principles of Knowledge, and of the Progress of Reason, from Sense to Science and Philosophy,* Vols. 1–2. Edinburgh: Strahan and Cadell, 1794.

———. "Observations on Granite." *Transactions of the Royal Society of Edinburgh* 3 (1794): 77–85.

———. *Theory of the Earth, with Proofs and Illustrations. 3* Vols.. Edinburgh: Cadell, Junior, Davies, and Creech, 1795.

———. "Theory of the Earth; or an Investigation of the Laws Observable in the Composition, Dissolution, and

Restoration of Land upon the Globe." *Transactions of the Royal Society of Edinburgh* 1, no. 2 (1788): 209–304.

Huxley, Julian. *Essays of a Biologist*. New York: Alfred A. Knopf, 1923. DOI: 10.5962/bhl.title.17941.

———. "Introduction." In Pierre Teilhard de Chardin, *The Phenomenon of Man*, 11–28. New York: HarperCollins, 2008.

Ingold, Tim. "Globes and Spheres: The Topology of Environmentalism." In *Environmentalism: The View from Anthropology*, edited by Kay Milton, 31–42. London: Routledge, 1993. DOI: 10.4324/9780203449653_chapter_2.

Jameson, Fredric. *Postmodernism, or, the Cultural Logic of Late Capitalism*. Durham: Duke University Press, 1991. DOI: 10.1215/9780822378419.

———. *The Seeds of Time*. New York: Columbia University Press, 1996.

Jonas, Hans. "Philosophy at the End of the Century: A Survey of Its Past and Future." *Social Research* 61, no. 4 (1994): 813–32.

Jones, Jean. "James Hutton's Agricultural Research and His Life as a Farmer." *Annals of Science* 42, no. 6 (1985): 573–601. DOI: 10.1080/00033798500200371.

Kant, Immanuel. *Critique of Judgement*. Edited by Nicholas Walker. Translated by James C. Meredith. Oxford: Oxford University Press, 2007.

———. *Critique of Pure Reason*. Edited and translated by Paul Guyer and Allen W. Wood. Cambridge: Cambridge University Press, 1998. DOI: 10.1017/CBO9780511804649.

Kaufmann, William J., and Larry L. Smarr. *Supercomputing and the Transformation of Science*. New York: Scientific American Library, 1993. DOI: 10.4267/2042/52011.

Keller, Evelyn F. "Organisms, Machines, and Thunderstorms: A History of Self-Organization, Part One." *Historical Studies in the Natural Sciences* 38, no. 1 (2008): 45–75. DOI: 10.1525/hsns.2008.38.1.45.

Kelly, Kevin. *Out of Control: The New Biology of Machines, Social Systems, and the Economic World*. Reading: Perseus Press, 1994.

King, Thomas M. *Teilhard de Chardin.* Wilmington: Glazier, 1988.
———. *Teilhard's Mysticism of Knowing.* New York: Seabury Press, 1981.
Kirchner, James W. "The Gaia Hypothesis: Can It Be Tested?" *Review of Geophysics* 27, no. 2 (1989): 223–35. DOI: 10.1029/RG027i002p00223.
Klingan, Katrin, and Christoph Rosol, eds. *Technosphäre: 100 Years of New Library.* Berlin: Haus der Kulturen der Welt, 2019.
Koyré, Alexandre. *From the Closed World to the Infinite Universe.* Baltimore: Johns Hopkins University Press, 1957.
Kulper, Amy C. "Architecture's Lapidarium: On the Lives of Geological Specimens." In *Architecture in the Anthropocene: Encounters among Design, Deep Time, Science and Philosophy,* edited by Etienne Turpin, 87–110. Ann Arbor: Open Humanities Press, 2013. DOI: 10.3998/ohp.12527215.0001.001.
Kwa, Chunglin. "Modelling Technologies of Control." *Science as Culture* 4, no. 3 (1994): 363–91. DOI: 10.1080/09505439409526393.
Lafontaine, Céline. "The Cybernetic Matrix of 'French Theory.'" *Theory, Culture & Society* 24, no. 5 (2007): 27–46. DOI: 10.1177/0263276407084637.
Lakatos, Imre. "History of Science and Its Rational Reconstructions." In *Scientific Revolutions,* edited by Ian Hacking, 107–27. Oxford: Oxford University Press, 1981.
Lamarck, Jean-Baptiste de. *Zoological Philosophy, or Exposition with Regard to the Natural History of Animals.* Translated by Hugh Elliot. New York: Hafner Publishing, 1963.
Lambright, W. Henry. "The Political Construction of Space Satellite Technology." *Science, Technology & Human Values* 19, no. 1 (1994): 47–69. DOI: 10.1177/016224399401900104.
Langmuir, Charles H., and Wallace S. Broecker. *How to Build a Habitable Planet: The Story of the Earth from the Big Bang to Humankind.* Princeton: Princeton University Press, 1985. DOI: 10.1515/9781400841974.

Lapenis, Andrei G. "Directed Evolution of the Biosphere: Biogeochemical Selection of Gaia." *The Professional Geographer* 54, no. 3 (2002): 379–91. DOI: 10.1111/0033-0124.00337.

Laplace, Pierre S. de. *A Philosophical Essay on Probabilities.* Translated by Frederick W. Truscott and Frederick L. Emory. New York: Wiley & Sons, 1951.

Lapo, Andrei V. *Traces of Bygone Biospheres.* Santa Fe: Synergetic Press, 1998.

Latour, Bruno. "A Cautious Prometheus? A Few Steps toward a Philosophy of Design with Special Attention to Peter Sloterdijk." In *In Medias Res: Peter Sloterdijk's Spherological Poetics of Being,* edited by Willem Schinkel and Liesbeth Noordegraaf-Eelens, 151–64. Amsterdam: Amsterdam University Press, 2011. DOI: 10.1515/9789048514502.

———. "Agency at the Time of the Anthropocene." *New Literary History* 45, no. 1 (2014): 1–18. DOI: 10.1353/nlh.2014.0003.

———. *Facing Gaia: Eight Lectures on the New Climatic Regime.* Translated by Catherine Porter. Cambridge: Polity Press, 2017.

———. "Love Your Monsters: Why We Must Care for Our Technologies as We Do Our Children." In *Love Your Monsters: Postenvironmentalism and the Anthropocene,* edited by Michael Schellenberger and Ted Nordhaus, 17–21. Oakland: The Breakthrough Institute, 2011.

———. *We Have Never Been Modern.* Translated by Catherine Porter. Cambridge: Harvard University Press, 1993.

Laudan, Rachel. *From Mineralogy to Geology: The Foundations of a Science, 1650–1830.* Chicago: University of Chicago Press, 1987. DOI: 10.7208/chicago/9780226924755.001.0001.

Lazier, Benjamin. "Earthrise; or, the Globalization of the World Picture." *American Historical Review* 116, no. 3 (2011): 602–30. DOI: 10.1086/ahr.116.3.602.

LeCain, Timothy J. "Heralding a New Humanism: The Radical Implications of Chakrabarty's 'Four Theses.'" In *Whose Anthropocene? Revisiting Dipesh Chakrabarty's "Four*

Theses," edited by Robert Emmett and Thomas Lekan, 15–20. Munich: Rachel Carson Center for Environment and Society, 2016. DOI: 10.5282/rcc/7421.

Lefebvre, Henri. "Dissolving City, Planetary Metamorphosis." Translated by Laurent Corroyer, Marianne Potvin, and Neil Brenner. *Environment and Planning D: Society and Space* 32, no. 2 (2004): 203–5. DOI: 10.1068/d3202tra.

Lemmens, Pieter, and Yuk Hui. "Reframing the Technosphere: Peter Sloterdijk and Bernard Stiegler's Anthropotechnological Diagnoses of the Anthropocene." *Krisis* 2 (2017): 26–41. https://archive.krisis.eu/reframing-the-technosphere-peter-sloterdijk-and-bernard-stieglers-anthropotechnologi-cal-diagnoses-of-the-anthropocene/.

Lettvin, Jerome Y., Warren S. McCulloch, Humberto R. Maturana, and Walter H. Pitts. "What the Frog's Eye Tells the Frog's Brain." *Proceedings of the Institute of Radio Engineers* 47 (1959): 1940–51. DOI: 10.1109/JRPROC.1959.287207.

Lewis, Simon L., and Mark A. Maslin. "Defining the Anthropocene." *Nature* 519, no. 7542 (2015): 171–80. DOI: 10.1038/nature14258.

Lidskog, Rolf, and Claire Waterton. "Anthropocene—A Cautious Welcome from Environmental Sociology?" *Environmental Sociology* 2, no. 4 (2016): 395–406. DOI: 10.1080/23251042.2016.1210841.

Lindberg, Susanna. "Onto-Technics in Bryant, Harman, and Nancy." *PhaenEx* (Winter 2018): 81–102. DOI: 10.22329/p.v12i2.5034. https://phaenex.uwindsor.ca/index.php/phaenex/article/view/5034.

Litfin, Karen T. *Ozone Discourses: Science and Politics in Global Environmental Cooperation.* New York: Columbia University Press, 1994.

Lövbrand, Eva, Silke Beck, Jason Chilvers, Tim Forsyth, Johan Hedrén, Mike Hulme, Rolf Lidskog, and Eleftheria Vasileiadou. "Who Speaks for the Future of Earth? How Critical Social Science Can Extend the Conversation on the

Anthropocene." *Global Environmental Change* 32 (2015): 211–18. DOI: 10.1016/j.gloenvcha.2015.03.012.

Lövbrand, Eva, Johannes Stripple, and Björn Wiman. "Earth System Governmentality: Reflections on Science in the Anthropocene." *Global Environmental Change* 19, no. 1 (2009): 7–13. DOI: 10.1016/j.gloenvcha.2008.10.002.

Lovejoy, Arthur O. *The Great Chain of Being: A Study of the History of an Idea.* Cambridge: Harvard University Press, 1936. DOI: 10.4159/9780674040335.

Lovelock, James E. "A Physical Basis for Life Detection Experiments." *Nature* 207, no. 4997 (1965): 568–70. DOI: 10.1038/207568a0.

———. *A Rough Ride to the Future.* Harmondsworth: Penguin, 2015.

———. *Gaia: A New Look at Life on Earth.* Oxford: Oxford University Press, 2000.

———. *Novacene: The Coming Age of Hyperintelligence.* Cambridge: MIT Press, 2019.

———. "Prehistory of Gaia." *New Scientist* 111 (1986): 51.

———. *The Ages of Gaia: A Biography of Our Living Earth.* New York: W.W. Norton, 1988.

———. *The Revenge of Gaia: Earth's Climate Crisis and the Fate of Humanity.* New York: Basic Books, 2007.

———. *The Vanishing Face of Gaia: A Final Warning.* New York: Basic Books, 2010.

Lovelock, James E., and C.E. Giffin. "Planetary Atmospheres: Compositional and Other Changes Associated with the Presence of Life." *Advances in the Astronautical Sciences* 25 (1969): 179–93.

Lovelock, James E., and Dian R. Hitchcock. "Life Detection by Atmospheric Analysis." *Icarus: International Journal of the Solar System* 7, no. 2 (1967): 149–59. DOI: 10.1016/0019-1035(67)90059-0.

Lovelock, James E., and Lynn Margulis. "The Atmosphere as Circulatory System of the Biosphere: The Gaia Hypothesis." *CoEvolution Quarterly* 6 (1975): 127–43.

———. "Atmospheric Homeostasis by and for the Biosphere: The Gaia Hypothesis." *Tellus* 26, nos. 1–2 (1974): 2–10. DOI: 10.3402/tellusa.v26i1-2.9731.

Luhmann, Niklas. "Deconstruction as Second-Order Observing." *New Literary History* 24, no. 4 (1993): 763–82. DOI: 10.2307/469391.

———. "Globalization or World Society: How to Conceive of Modern Society?" *International Review of Sociology* 7, no. 1 (1997): 67–79. DOI: 10.1080/03906701.1997.9971223.

———. "The Cognitive Program of Constructivism and a Reality that Remains Unknown." In *Theories of Distinction: Redescribing the Descriptions of Modernity,* edited by William Rasch, translated by Joseph O'Neil, Elliot Schreiber, Kerstin Behnke, and William Whobrey, 128–53. Stanford: Stanford University Press, 2002. DOI: 10.1515/9781503619340-008.

Luke, Timothy W. "On Environmentality: Geo-Power and Eco-Knowledge in the Discourses of Contemporary Environmentalism." *Cultural Critique* 31 (1995): 57–81. DOI: 10.2307/1354445.

Lyell, Charles. *Principles of Geology: Being an Inquiry How Far the Former Changes of the Earth's Surface Are Referable to Causes Now in Operation.* Vol. 1. London: John Murray, 1835. DOI: 10.5962/bhl.title.50199.

Macfie, Alec. "The Invisible Hand of Jupiter." *Journal of the History of Ideas* 32, no. 4 (1971): 595–99. DOI: 10.2307/2708980.

Malabou, Catherine. "The Brain of History, or, the Mentality of the Anthropocene." *South Atlantic Quarterly* 116, no. 1 (2017): 39–53. DOI: 10.1215/00382876-3749304.

Malm, Andreas. "Against Hybridism: Why We Need to Distinguish between Nature and Society, Now More Than Ever." *Historical Materialism* 27, no. 2 (2019): 156–87. DOI: 10.1163/1569206X-00001610.

———. *Fossil Capital: The Rise of Steam Power and the Roots of Global Warming.* London: Verso, 2016.

———. *The Progress of This Storm: Nature and Society in a Warming World.* London: Verso, 2017.
Malm, Andreas, and Alf Hornborg. "The Geology of Mankind? A Critique of the Anthropocene Narrative." *Anthropocene Review* 1, no. 1 (2014): 62–69. DOI: 10.1177/2053019613516291.
Margulis, Lynn. "Big Trouble in Biology: Physiological Autopoiesis versus Mechanistic Neo-Darwinism." In *Slanted Truths: Essays on Gaia, Symbiosis, and Evolution,* edited by Lynn Margulis and Dorion Sagan, 265–82. Berlin: Springer, 1997. DOI: 10.1007/978-1-4612-2284-2_20.
———. "Foreword to the English-language Edition." In Vladimir I. Vernadsky, *The Biosphere,* 14–19. New York: Copernicus, 1998. DOI: 10.1007/978-1-4612-1750-3.
———. "Gaia Is a Tough Bitch." In *The Third Culture: Beyond the Scientific Revolution,* edited by John Brockman, 129–51. New York: Simon & Schuster, 1995.
———. "Kingdom Animalia: The Zoological Malaise from a Microbial Perspective." *American Zoologist* 30, no. 4 (1990): 861–75. DOI: 10.1093/icb/30.4.861.
———. *Origin of Eukaryotic Cells: Evidence and Research Implications for a Theory of the Origin and Evolution of Microbial, Plant, and Animal Cells on the Precambrian Earth.* New Haven: Yale University Press, 1970.
———. "Symbiogenesis: A New Principle of Evolution Rediscovery of Boris Mikhaylovich Kozo-Polyansky (1890–1957)." *Paleontological Journal* 44, no. 12 (2010): 1525–39. DOI: 10.1134/S0031030110120087.
———. *Symbiotic Planet: A New Look at Evolution.* New York: Basic Books, 1998.
Margulis, Lynn, and Dorion Sagan. *Acquiring Genomes: A Theory of the Origin of Species.* New York: Basic Books, 2003.
———. *Microcosmos: Four Billion Years of Evolution from Our Microbial Ancestors.* Berkeley: University of California Press, 1997. DOI: 10.1525/9780520340510.
———. *Origins of Sex: Three Billion Years of Genetic Recombination.* New Haven: Yale University Press, 1986.

———. *What Is Life?* Berkeley: University of California Press, 1995.

Margulis, Lynn, and James E. Lovelock. "Gaia and Geognosy." In *Global Ecology: Towards a Science of the Biosphere,* edited by Mitchell B. Rambler, Lynn Margulis, and René Fester, 1–30. San Diego: Academic Press, 1989.

Marx, Karl. *Grundrisse: Foundations of the Critique of Political Economy.* Translated by Martin Nicolaus. Harmondsworth: Penguin, 1973.

———. "Letter to Friedrich Engels, 18 June 1862." Translated by Peter Ross and Betty Ross. In *The Collected Works of Karl Marx and Friedrich Engels,* Vol. 41: *Letters* 1860–1864, edited by Jack Cohen, Maurice Cornforth, Maurice Dobb, Eric J. Hobsbawm, James Klugmann, and Margaret Mynatt, 380–82. London: Lawrence and Wishart, 2005.

Marx, Leo. "The Idea of 'Technology' and Postmodern Pessimism." In *Does Technology Drive History? The Dilemma of Technological Determinism,* edited by Merritt R. Smith and Leo Marx, 237–58. Cambridge: MIT Press, 1994.

———. *The Machine in the Garden: Technology and the Pastoral Ideal in America.* Oxford: Oxford University Press, 2000.

Maturana, Humberto R., and Francisco J. Varela. *Autopoiesis and Cognition: The Realization of the Living.* Dordrecht: D. Reidel, 1980. DOI: 10.1007/978-94-009-8947-4.

———. *The Tree of Knowledge: The Biological Roots of Human Understanding.* Translated by Robert Paolucci. Boston: Shambhala, 1987.

McKibben, Bill. *The End of Nature.* New York: Random House, 1989.

McLuhan, Marshall. "At the Moment of Sputnik the Planet Became a Global Theater in Which There Are No Spectators but Only Actors." *Journal of Communication* 24, no. 1 (1974), 48–58. DOI: 10.1111/j.1460-2466.1974.tb00354.x.

McNeill, William H. "Passing Strange: The Convergence of Evolutionary Science with Scientific History." *History and Theory* 40, no. 1 (2001): 1–15. DOI: 10.1111/0018-2656.00149.

Medawar, Peter B. "VI.—Critical Notice. *The Phenomenon of Man*. By Pierre Teilhard de Chardin. With an introduction by Sir Julian Huxley. Collins, London, 1959. 25s." *Mind* 70, no. 277 (1961): 99–106. DOI: 10.1093/mind/LXX.277.99.

Melitopoulos, Angela, and Maurizio Lazzarato. "Machinic Animism." *Deleuze Studies* 6, no. 2 (2012): 240–49. DOI: 10.3366/dls.2012.0060.

Millennium Ecosystem Assessment (MEA). *Ecosystems and Human Well-Being: Current State and Trends*, Vol. 1. Washington, DC: Island Press, 2005.

———. *Living beyond Our Means: Natural Assets and Human Well-Being. Statement from the Board*. New York: MEA, 2005.

Mitcham, Carl. *Thinking through Technology: The Path between Engineering and Philosophy*. Chicago: University of Chicago Press, 1994. DOI: 10.7208/chicago/9780226825397.001.0001.

Monastersky, Richard. "Anthropocene: The Human Age." *Nature* 519, no. 7542 (2015): 144–47. DOI: 10.1038/519144a.

Monod, Jacques. *Chance and Necessity: An Essay on the Natural Philosophy of Modern Biology*. Translated by Austryn Wainhouse. New York: Vintage Books, 1972.

———. "From Molecular Biology to the Ethics of Knowledge." Translated by Arnold Pomerans. *The Human Context* 1, no. 4 (1969): 325–36.

Moore, Jason W. "Anthropocene or Capitalocene? Nature, History, and the Crisis of Capitalism." In *Anthropocene or Capitalocene? Nature, History, and the Crisis of Capitalism*, edited by Jason W. Moore, 1–11. Oakland: Kairos, 2016.

———. *Capitalism in the Web of Life: Ecology and the Accumulation of Capital*. London: Verso, 2015.

———. "Toward a Singular Metabolism: Epistemic Rifts and Environment-Making in the Capitalist World-Ecology." In *New Geographies*, Vol. 6: *Grounding Metabolism*, edited by Daniel Ibañez and Nikos Katsikis, 10–19. Cambridge: Harvard University Press, 2014.

———. "Transcending the Metabolic Rift: A Theory of Crises in the Capitalist World-Ecology." *Journal of Peasant Studies* 38, no. 1 (2011): 1–46. DOI: 10.1080/03066150.2010.538579.

Morin, Marie-Eve. "Cohabitating in the Globalised World: Peter Sloterdijk's Global Foams and Bruno Latour's Cosmopolitics." *Environment and Planning D: Society and Space* 27, no. 1 (2009): 58–72. DOI: 10.1068/d4908.

Morton, Timothy. *Dark Ecology: For a Logic of Future Coexistence.* New York: Columbia University Press, 2016. DOI: 10.7312/mort17752.

———. *Hyperobjects: Philosophy and Ecology after the End of the World.* Minneapolis: University of Minnesota Press, 2013.

———. *The Ecological Thought.* Cambridge: Harvard University Press, 2010. DOI: 10.4159/9780674056732.

Moynihan, Thomas. *Spinal Catastrophism: A Secret History.* Falmouth: Urbanomic, 2019.

———. *X-Risk: How Humanity Discovered Its Own Extinction.* Falmouth: Urbanomic, 2020.

Mumford, Lewis. *Technics and Civilization.* London: Routledge & Kegan Paul, 1955.

———. *Technics and Human Development,* Vol. 1: *The Myth of the Machine.* New York: Harcourt, 1967.

Nancy, Jean-Luc. *Corpus.* Translated by Richard A. Rand. New York: Fordham University Press, 2008. DOI: 10.2307/j.ctt13x04c6.

Nancy, Jean-Luc, and Aurelien Barrau. *What's These Worlds Coming To?* Translated by Travis Holloway and Flor Méchain. New York: Fordham University Press, 2015. DOI: 10.5422/fordham/9780823263332.001.0001.

National Research Council (NRC). *Human Dimensions of Global Environmental Change: Research Pathways for the Next Decade.* Washington, DC: National Academy Press, 1999.

Neubert, Christoph, and Serjoscha Wiemer. "Rewriting the Matrix of Life: Biomedia Between Ecological Crisis and Playful Actions." *communications +1* 3, no. 1 (2014): 1–31. DOI: 10.7275/R50V89RS.

Odum, Eugene P. *Fundamentals of Ecology.* Philadelphia: W.B. Saunders, 1959.

Odum, Howard T. *Environment, Power, and Society.* New York: Wiley, 1970.

Oldfield, Frank, Anthony D. Barnosky, John Dearing, Marina Fischer-Kowalski, John R. McNeill, Will Steffen, and Jan A. Zalasiewicz. "The Anthropocene Review: Its Significance, Implications and the Rationale for a New Transdisciplinary Journal." *Anthropocene Review* 1, no. 1 (2013): 3–7. DOI: 10.1177/2053019613500445.

Oldfield, Jonathan D., and Denis J.B. Shaw. "A Russian Geographical Tradition? The Contested Canon of Russian and Soviet Geography, 1884–1953." *Journal of Historical Geography* 49 (2015): 75–84. DOI: 10.1016/j.jhg.2015.04.015.

———. "V.I. Vernadsky and the Noosphere Concept: Russian Understandings of Society-Nature Interaction." *Geoforum* 37, no. 1 (2006): 145–54. DOI: 10.1016/j.geoforum.2005.01.004.

Oldroyd, David R. *The Highlands Controversy: Constructing Geological Knowledge through Fieldwork in Nineteenth-Century Britain.* Chicago: University of Chicago Press, 1990.

———. *Thinking about the Earth: A History of Ideas in Geology.* Cambridge: Harvard University Press, 1996.

Osborne, Thomas. "Vitalism as Pathos." *Biosemiotics* 9 (2016): 185–205. DOI: 10.1007/s12304-016-9254-7.

Parikka, Jussi. *A Geology of Media.* Minneapolis: University of Minnesota Press, 2015. DOI: 10.5749/minnesota/9780816695515.001.0001.

Pask, Gordon. "Introduction: Different Kinds of Cybernetics." In *New Perspectives on Cybernetics: Self-Organization, Autonomy, and Connectionism,* edited by Gertrudis Van de Vijver, 11–31. Dordrecht: Kluwer, 1992. DOI: 10.1007/978-94-015-8062-5_1.

Pellizzoni, Luigi. "New Materialism and Runaway Capitalism: A Critical Assessment." *Soft Power* 5, no. 1 (2017): 63–80. DOI: 1017450/170104.

———. *Ontological Politics in a Disposable World: The New Mastery of Nature.* London: Routledge, 2015. DOI: 10.4324/9781315598925.

Pimm, Stuart L., Clinton N. Jenkins, Robin Abell, Thomas M. Brooks, J.L. Gittleman, Lucas N. Joppa, P.H. Raven, Callum M. Roberts, and Joseph C. Sexton. "The Biodiversity of Species and Their Rates of Extinction, Distribution, and Protection." *Science* 344, no. 6187 (2014): 1–10. DOI: 10.1126/science.1246752.

Piqueras, Mercè. "Meeting the Biospheres: On the Translations of Vernadsky's Work." *International Microbiology* 1, no. 2 (1998): 165–70.

Playfair, John. "Biographical Account of the Late Dr. James Hutton." *Transactions of the Royal Society of Edinburgh* 5 (1805): 39–99. DOI: 10.1017/S0263593300090039.

———. *Illustrations of the Huttonian Theory of the Earth.* Cambridge: Cambridge University Press, 2011. DOI: 10.1017/CBO9780511973086.

Polunin, Nicholas, and Jacques Grinevald. "Vernadsky and Biospheral Ecology." *Environmental Conservation* 15, no. 2 (1988): 117–22. DOI: 10.1017/S0376892900028915.

Porter, Roy. "Gentlemen and Geology: The Emergence of a Scientific Career, 1660–1920." *The Historical Journal* 21, no. 4 (1978): 809–36. DOI: 10.1017/S0018246X78000024.

Pronk, Jan. "The Amsterdam Declaration on Global Change." In *Challenges of a Changing Earth. Global Change — The IGBP Series,* edited by Will Steffen, Jill Jäger, David J. Carson, and Clare Bradshaw, 207–8. Berlin: Springer, 2002. DOI: 10.1007/978-3-642-19016-2_40.

Ray, Gene. "Terror and the Sublime in the So-Called Anthropocene." *Liminalities: A Journal of Performance Studies* 16, no. 2 (2020): 1–20.

Renn, Jürgen. "The Evolution of Knowledge: Rethinking Science in the Anthropocene." *Journal of History of Science and Technology* 12, no. 1 (2018): 1–22. DOI: 10.2478/host-2018-0001.

Ricciardi, Anthony. "Are Modern Biological Invasions an Unprecedented Form of Global Change?" *Conservation Biology* 21, no. 2 (2007): 329–36. DOI: 10.1111/j.1523-1739.2006.00615.x.

Richardson, Lewis F. *Weather Prediction by Numerical Process.* Cambridge: Cambridge University Press, 1922. DOI: 10.1017/CBO9780511618291.

Rockström, Johan, Will Steffen, Kevin Noone, Åsa Persson, F. Stuart III Chapin, Eric Lambin, Timothy M. Lenton, Marten Scheffer, Carl Folke, and Hans J. Schellnhuber. "Planetary Boundaries: Exploring the Safe Operating Space for Humanity." *Ecology and Society* 14, no. 2 (2009): art. 32. DOI: 10.5751/ES-03180-140232.

Rosen, Robert. "On a Logical Paradox Implicit in the Notion of a Self-Reproducing Automata." *Bulletin of Mathematical Biophysics* 21 (1959): 387–94. DOI: 10.1007/BF02477897.

Roudeau, Cécile. "The Buried Scales of Deep Time: Beneath the Nation, beyond the Human … and Back?" *Transatlantica* 1 (2015): 1–14. DOI: 10.4000/transatlantica.7455.

Rubey, William W. "Geologic History of Sea Water: An Attempt to State the Problem." *Geological Society of America Bulletin* 62, no. 9 (1951): 1111–48. DOI: 10.1130/0016-7606(1951)62[1111:GHOSW]2.0.CO;2.

Rudwick, Martin J. S. *Bursting the Limits of Time: The Reconstruction of Geohistory in the Age of Revolution.* Chicago: University of Chicago Press, 2005. DOI: 10.7208/chicago/9780226731148.001.0001.

———. *Earth's Deep History: How It Was Discovered and Why It Matters.* Chicago: University of Chicago Press, 2014. DOI: 10.7208/chicago/9780226204093.001.0001.

———. *Georges Cuvier, Fossil Bones, and Geological Catastrophes: New Translations and Interpretations of the Primary Texts.* Chicago: University of Chicago Press, 2008. DOI: 10.7208/chicago/9780226731087.001.0001.

———. *The Great Devonian Controversy: The Shaping of Scientific Knowledge among Gentlemanly Specialists.* Chicago: University of Chicago Press, 1985. DOI: 10.7208/chicago/9780226731001.001.0001.

Rutherford, Paul. "The Entry of Life into History." In *Discourses of the Environment,* edited by Éric Darier, 37–62. Oxford: Blackwell, 1999.

Rutsky, R.L. *High Technē: Art and Technology from the Machine Aesthetic to the Posthuman.* Minneapolis: University of Minnesota Press, 1999.

Sagan, Dorion. "Möbius Trip: The Technosphere and Our Science Fiction Reality." *Technosphere Magazine,* November 15, 2016. https://www.anthropocene-curriculum.org/contribution/mobius-trip-the-technosphere-and-our-science-fiction-reality/.

Sagan, Dorion, and Jessica H. Whiteside. "Gradient-Reduction Theory: Thermodynamics and the Purpose of Life." In *Scientists Debate Gaia: The Next Century,* edited by Stephen H. Schneider, James R. Miller, Eileen Crist, and Penelope J. Boston, 173–86. Cambridge: MIT Press, 2004. DOI: 10.7551/mitpress/9780262194983.003.0017.

Sagan, Dorion, and Lynn Margulis. "Gaia and the Evolution of Machines." *Whole Earth Review* 55 (1987): 15–21.

———. "Welcome to the Machine." In *Dazzle Gradually: Reflections on the Nature of Nature,* edited by Lynn Margulis and Dorion Sagan, 76–88. White River Junction: Chelsea Green Publishing, 2007.

Saldanha, Arun. "Mechanosphere: Man, Earth, Capital." In *Deleuze and the Non/Human,* edited by Jon Roffe and Hannah Stark, 197–216. Berlin: Springer, 2015. DOI: 10.1057/9781137453693_12.

Saldanha, Arun, and Hannah Stark. "A New Earth: Deleuze and Guattari in the Anthropocene." *Deleuze Studies* 10, no. 4 (2016): 427–39. DOI: 10.3366/dls.2016.0237.

Schaffer, Simon. "Babbage's Intelligence: Calculating Engines and the Factory System." *Critical Inquiry* 21, no. 1 (1994): 203–27. DOI: 10.1086/448746.

Schelling, Friedrich W.J. von. *Ideas for a Philosophy of Nature.* Edited and translated by Errol E. Harris and Peter Heath. Cambridge: Cambridge University Press, 1988.

———. *On University Studies.* Translated by E. S. Morgan. Athens: Ohio University Press, 1966.

———. *System of Transcendental Idealism.* Translated by Peter Heath. Charlottesville: University of Virginia Press, 1978.

Schellnhuber, Hans J. "'Earth System' Analysis and the Second Copernican Revolution." *Nature* 402, no. 6761 (1999): C19–C23. DOI: 10.1038/35011515.

Scherer, Bernd M. "A Report: An Introduction." In *The Anthropocene Project: A Report,* edited by Bernd M. Scherer, 4–11. Berlin: Haus der Kulturen der Welt, 2014.

Schopenhauer, Arthur. *The World as Will and Representation.* Vol. 2. Translated by Eric F.J. Payne. New York: Dover, 1969.

Schumpeter, Joseph. *Capitalism, Socialism, and Democracy.* London: Routledge, 1994. DOI: 10.4324/9780203857090.

Searles, Harold. "Unconscious Processes in Relation to the Environmental Crisis." *Psychoanalytic Review* 59, no. 3 (1972): 361–74.

Secord, James A. *Controversy in Victorian Geology: The Cambrian–Silurian Dispute.* Princeton: Princeton University Press, 2014. DOI: 10.1515/9781400854660.

Şengör, A.M. Celâl. *Is the Present the Key to the Past or Is the Past the Key to the Present? James Hutton and Adam Smith versus Abraham Gottlob Werner and Karl Marx in Interpreting History.* The Geological Society of America Special Paper, Vol. 355. Boulder: The Geological Society of America, 2001. DOI: 10.1130/0-8137-2355-8.1.

Serres, Michel. "Turner Translates Carnot." In *Hermes: Literature, Science, Philosophy,* edited by Josué V. Harari and David F. Bell, 54–62. Baltimore: Johns Hopkins University Press, 1982.

Sloterdijk, Peter. *Bubbles—Spheres,* Vol. 1: *Microspherology.* Translated by Wieland Hoban. Los Angeles: Semiotext(e), 2011.

———. *Globes—Spheres,* Vol. 2: *Macrosphereology.* Translated by Wieland Hoban. Los Angeles: Semiotext(e), 2014.

———. "*Rules for the Human Zoo:* A Response to the *Letter on Humanism.*" Translated by Mary V. Rorty. *Environment and*

Planning D: Society and Space 27, no. 1 (2007): 12–28. DOI: 10.1068/dst3.

———. *Terror from the Air.* Translated by Amy Patton and Steve Corcoran. Cambridge: MIT Press, 2002.

Smil, Václav. *The Earth's Biosphere: Evolution, Dynamics, and Change.* Cambridge: MIT Press, 2002. DOI: 10.7551/mitpress/2551.001.0001.

Smith, Adam. *An Inquiry into the Nature and Causes of the Wealth of Nations.* Edited by Roy H. Campbell, Andrew S. Skinner, and William B. Todd. 4 Vols.. Oxford: Oxford University Press, 1976. DOI: 10.1093/actrade/9780199269563.book.1.

———. *Essays on Philosophical Subjects.* Edited by William P.D. Wightman, John C. Bryce, and Ian S. Ross. Oxford: Oxford University Press, 1980. DOI: 10.1093/oseo/instance.00042833.

Smith, Bruce D., and Melinda A. Zeder. "The Onset of the Anthropocene." *Anthropocene* 4 (2013): 8–13. DOI: 10.1016/j.ancene.2013.05.001.

Steffen, Will, Wendy Broadgate, Lisa Deutsch, Owen Gaffney, and Cornelia Ludwig. "The Trajectory of the Anthropocene: The Great Acceleration." *Anthropocene Review* 2, no. 1 (2015): 81–98. DOI: 10.1177/2053019614564785.

Steffen, Will, Paul J. Crutzen, and John R. McNeill. "The Anthropocene: Are Humans Now Overwhelming the Great Forces of Nature?" *Ambio* 36, no. 8 (2007): 614–21. DOI: 10.1579/0044-7447(2007)36%5B614:TAAHNO%5D2.0.CO;2.

Steffen, Will, Jacques Grinevald, Paul J. Crutzen, and John R. McNeill. "The Anthropocene: Conceptual and Historical Perspectives." *Philosophical Transactions of the Royal Society A* 369, no. 1938 (2011): 842–67. DOI: 10.1098/rsta.2010.0327.

Steffen, Will, Åsa Persson, Lisa Deutsch, Jan A. Zalasiewicz, Mark Williams, Katherine Richardson, Carole Crumley, Paul J. Crutzen, Carl Folke, Line Gordon, Mario Molina, Veerabhadran Ramanathan, Johan Rockström, Marten Scheffer, Hans Joachim Schellnhuber, and Uno Svedin. "The Anthropocene: From Global Change to Planetary

Stewardship." *Ambio* 40, no. 7 (2011): 739–61. DOI: 10.1007/s13280-011-0185-x.

Steffen, Will, Angelina Sanderson, Peter Tyson, Jill Jäger, Pamela Matson, Berrien Moore, Frank Oldfield, Katherine Richardson, Hans J. Schellnhuber, B.L. Turner, and Robert J. Wasson, eds. *Global Change and the Earth System: A Planet under Pressure — IGBP Global Change Series.* Berlin: Springer, 2005. DOI: 10.1007/b137870.

Stiegler, Bernard. *Technics and Time,* Vol. 1: *The Fault of Epimetheus.* Translated by George Colins and Richard Beardsworth. Stanford: Stanford University Press, 1998. DOI: 10.1515/9781503616738.

Stone, Alison. *Petrified Intelligence: Nature in Hegel's Philosophy.* Albany: SUNY Press, 2005.

Stoppani, Antonio. "First Period of the Anthropozoic Era." Edited and translated by Valeria Federighi and Etienne Turpin. In *Making the Geologic Now: Responses to the Material Conditions of Contemporary Life,* edited by Elizabeth Ellsworth and Jamie Kruse, 36–41. Brooklyn: punctum books, 2013. DOI: 10.21983/P3.0014.1.00.

Suess, Eduard. *Die Entstehung der Alpen.* Vienna: W. Braunmiller, 1875.

Szerszynski, Bronislaw. "The End of the End of Nature: The Anthropocene and the Fate of the Human." *Oxford Literary Review* 34, no. 2 (2012): 165–84. DOI: 10.3366/olr.2012.0040.

———. "Out of the Metazoic? Animals as a Transitional Form in Planetary Evolution." In *Thinking about Animals in the Age of the Anthropocene,* edited by Morten Tønnesen, Kristin Armstrong Oma, and Silver Rattasepp, 163–79. Lanham: Rowman & Littlefield, 2016.

———. "Planetary Mobilities: Movement, Memory, and Emergence in the Body of the Earth." *Mobilities* 11, no. 4 (2016): 614–28. DOI: 10.1080/17450101.2016.1211828.

Sörlin, Sverker, and Nina Wormbs. "Environing Technologies: A Theory of Making Environment." *History and Technology* 34, no. 2 (2018): 101–25. DOI: 10.1080/07341512.2018.1548066.

Teilhard de Chardin, Pierre. *Activation of Energy: Enlightening Reflections on Spiritual Energy.* New York: HarperCollins, 1978.
———. *Christianity and Evolution.* New York: Harvest, 1974.
———. *Letters from a Traveler.* New York: Harper & Row, 1962.
———. *Man's Place in Nature.* Translated by René Hague. New York: Harper & Row, 1966.
———. *The Future of Man.* New York: HarperCollins, 2008.
———. *The Heart of Matter.* Translated by René Hague. New York: Harcourt, 1979.
———. *The Phenomenon of Man.* Translated by Bernard Wall. New York: HarperCollins, 2008.
———. *The Vision of the Past.* Translated by John M. Cohen. New York: Harper & Row, 1966.
———. *Writings in Time of War.* New York: HarperCollins, 1968.
Thompson, Evan. *Mind in Life: Biology, Phenomenology, and the Sciences of Mind.* Cambridge: Harvard University Press, 2007.
Thompson, William I. *Coming into Being: Artifacts and Texts in the Evolution of Consciousness.* New York: St. Martin's Griffin, 1998.
———. *Pacific Shift.* San Francisco: Sierra Club, 1985.
Thomson, Iain D. "From the Question Concerning Technology to the Quest for a Democratic Technology: Heidegger, Marcuse, Feenberg." *Inquiry: An Interdisciplinary Journal of Philosophy* 43, no. 2 (2000): 203–16. DOI: 10.1080/002017400407753.
———. "What's Wrong with Being a Technological Essentialist? A Response to Feenberg." *Inquiry: An Interdisciplinary Journal of Philosophy* 43, no. 4 (2000): 429–44. DOI: 10.1080/002017400750051233.
Turner, Fred. *From Counterculture to Cyberculture: Stewart Brand, the Whole Earth Network, and the Rise of Digital Utopianism.* Chicago: University of Chicago Press, 2006. DOI: 10.7208/chicago/9780226817439.001.0001.

Uhrqvist, Ola. "Seeing and Knowing the Earth as a System: An Effective History of Global Environmental Change Research as Scientific and Political Practice." PhD diss., Linköping University, 2014. DOI: 10.3384/diss.diva-110654.

Uhrqvist, Ola, and Eva Lövbrand. "Rendering Global Change Problematic: The Constitutive Effects of Earth System Research in the IGBP and the IHDP." *Environmental Politics* 23, no. 2 (2014): 339–56. DOI: 10.1080/09644016.2013.835964.

Van Vuuren, Detlef P. *Growing within Limits: A Report to the Global Assembly 2009 of the Club of Rome.* Bilthoven: PBL Netherlands Environmental Assessment Agency, 2009.

Vatican II. *The Pastoral Constitution on the Church in the Modern World/Gaudium et spes.* Rome: The Vatican, 1965.

Vernadsky, Vladimir I. *Geochemistry and the Biosphere: Essays by Vladimir I. Vernadsky.* Edited by Frank B. Salisbury. Translated by Olga Barash. Santa Fe: Synergetic Press, 2007.

———. "'Problems of Biogeochemistry' II: The Fundamental Matter-Energy Difference between the Living and Inert Natural Bodies of the Biosphere." Edited by G. Evelyn Hutchinson. Translated by George Vernadsky. *Transactions of the Connecticut Academy of Arts and Sciences* 35 (1944): 483–517.

———. *The Biosphere.* Translated by David B. Langmuir. New York: Copernicus, 1998. DOI: 10.1007/978-1-4612-1750-3.

———. "The Biosphere and the Noösphere." *American Scientist* 33, no. 1 (1945): 1–12. https://www.jstor.org/stable/27826043.

———. "The Transition from the Biosphere to the Noösphere: Excerpts from Scientific Thought as a Planetary Phenomenon, 1938." Translated by William Jones. *Twenty-First Century Science & Technology* (Spring–Summer 2012): 10–31.

Vitousek, Peter M., Carla M. D'Antonio, Lloyd L. Loope, Marcel Rejmánek, and Randy Westbrooks. "Introduced Species: A Significant Component of Human-Caused Global Change." *New Zealand Journal of Ecology* 21, no. 1 (1997): 1–16.

Vitousek, Peter M., Harold A. Mooney, Jane Lubchenco, and Jerry M. Melillo. "Human Domination of Earth's Ecosystems." *Science* 277, no. 5325 (1997): 494–99. DOI: 10.1126/science.277.5325.494.

Wapner, Paul. *Living through the End of Nature: The Future of American Environmentalism.* Cambridge: MIT Press, 2010. DOI: 10.7551/mitpress/8454.001.0001.

Weber, Max. *The Sociology of Religion.* Translated by Ephraim Fischoff. London: Methuen, 1965.

Weber, Samuel. "Upsetting the Setup: Remarks on Heidegger's Questing after Technics." In *Mass Mediauras: Form, Technics, Media,* edited by Alan Cholodenko, 55–75. Stanford: Stanford University Press, 1996.

Wertime, Theodore A. "Pyrotechnology: Man's First Industrial Uses of Fire." *American Scientist* 61, no. 6 (1973): 670–82. https://www.jstor.org/stable/27844070.

Whewell, William. *History of the Inductive Sciences from the Earliest to the Present Times.* Vol. 3. London: John W. Parker, 1837. DOI: 10.5962/bhl.title.31574.

White, Damian F., and Chris Wilbert, eds. *Technonatures: Environments, Technologies, Spaces, and Places in the Twenty-First Century.* Waterloo: Wilfrid Laurier University Press, 2010.

Wiener, Norbert. *Cybernetics: Or Control and Communication in the Animal and the Machine.* Cambridge: MIT Press, 1948. DOI: 10.7551/mitpress/11810.001.0001.

Winner, Langdon. *Autonomous Technology: Technics-out-of-Control as a Theme in Political Thought.* Cambridge: MIT Press, 1977.

Withers, Charles W.J. "On Georgics and Geology: James Hutton's 'Elements of Agriculture' and Agricultural Science in Eighteenth-Century Scotland." *Agricultural History Review* 42, no. 1 (1992): 38–48.

Wolfe, Cary. "In Search of Posthumanist Theory: The Second-Order Cybernetics of Maturana and Varela." In *Observing Complexity: Systems Theory and Postmodernity,* edited

by Cary Wolfe and William Rasch, 163–96. Minneapolis: University of Minnesota Press, 2000.
———. *What Is Posthumanism?* Minneapolis: University of Minnesota Press, 2009.
Wolfendale, Peter. "The Reformatting of Homo Sapiens." *Angelaki: Journal of the Theoretical Humanities* 24, no. 1 (2019): 55–66. DOI: 10.1080/0969725X.2019.1568733.
Wood, Linda S. *A More Perfect Union: Holistic Worldviews and the Transformation of American Culture after World War II.* Oxford: Oxford University Press, 2010. DOI: 10.1093/acprof:oso/9780195377743.001.0001.
Zalasiewicz, Jan A., Sverker Sörlin, Libby Robin, and Jacques Grinevald. "Introduction: Buffon and the History of the Earth." In Georges-Louis Leclerc, *The Epochs of Nature,* edited and translated by Jan A. Zalasiewicz, Ann-Sophie Milon, and Mateusz Zalasiewicz, xiii–xxxiv. Chicago: University of Chicago Press, 2018. DOI: 10.7208/chicago/9780226395579.001.0001.
Zalasiewicz, Jan A., Will Steffen, Reinhold Leinfelder, Mark Williams, and Colin Waters. "Petrifying Earth Processes: The Stratigraphic Imprint of Key Earth System Parameters in the Anthropocene." *Theory, Culture & Society* 34, nos. 2–3 (2017): 83–104. DOI: 10.1177/0263276417690587.
Zalasiewicz, Jan A., Mark Williams, Will Steffen, and Paul J. Crutzen. "The New World of the Anthropocene." *Environmental Science and Technology* 44, no. 7 (2010): 2228–31. DOI: 10.1021/es903118j.
Zalasiewicz, Jan A., Mark Williams, Colin N. Waters, Anthony D. Barnosky, John Palmesino, Ann-Sofi Rönnskog, Matt Edgeworth, Cath Neal, Alejandro Cearreta, Erle C. Ellis, et al. "Scale and Diversity of the Physical Technosphere: A Geological Perspective." *Anthropocene Review* 4, no. 1 (2017): 9–22. DOI: 10.1177/2053019616677743.
Zielinski, Siegfried. *Deep Time of the Media: Toward an Archaeology of Hearing and Seeing by Technical Means.* Translated by Gloria Custance. Cambridge: MIT Press, 2006.

Ziolkowski, Theodore. *German Romanticism and Its Institutions.* Princeton: Princeton University Press, 1990. DOI: 10.1515/9780691225760.

Žižek, Slavoj. *Living in the End Times.* London: Verso, 2010.

———. *The Indivisible Remainder: On Schelling and Related Matters.* London: Verso, 1996.

Zwier, Jochem, and Vincent Blok. "Saving Earth: Encountering Heidegger's Philosophy of Technology in the Anthropocene." *Techné: Research in Philosophy and Technology* 21, nos. 2–3 (2017): 222–42. DOI: 10.5840/techne201772167.

———. "Seeing through the Fumes: Technology and Asymmetry in the Anthropocene." *Human Studies* 42 (2019): 621–46. DOI: 10.1007/s10746-019-09508-4.

Made in the USA
Middletown, DE
28 October 2023

41526904R00195